TRAVEL
YOUR WAY

Nathan James Thomas founded Intrepid Times in 2014 as a vehicle for sharing stories from the road and as an excuse to meet and interview his favourite writers. It has since grown into a popular home for travel writing with heart, attracting hundreds of contributors and thousands of readers from around the world, and selected as a finalist for Consumer Publication of the Year (online) in the 2021 Travel Media Awards. Nathan's own travel writing has been published in places like *Roads and Kingdoms, Outpost Magazine* and *New Zealand Memories*, and his work as a ghostwriter has been featured in *Forbes*, the *Huffington Post* and the *Harvard Business Review*. Originally from New Zealand, Nathan lived in China for two years and is currently based in Eastern Europe. He is the co-editor of the travel writing compilation *Fearless Footsteps*, also published by Exisle.

TRAVEL YOUR WAY

Rediscover the world, on your own terms

NATHAN JAMES THOMAS

First published 2022

Exisle Publishing Pty Ltd
PO Box 864, Chatswood, NSW 2057, Australia
226 High Street, Dunedin, 9016, New Zealand
www.exislepublishing.com

A CiP record for this book is available from the
National Library of Australia.

ISBN 978-1-925820-58-4

Designed by Shaun Jury
Typeset in Scala, Sagona and Din
Printed and bound in Great Britain by Clays Ltd, Elcograf
S.p.A.

This book uses paper sourced under ISO 14001 guidelines
from well-managed forests and other controlled sources.

10 9 8 7 6 5 4 3 2 1

Disclaimer
While this book is intended as a general information
resource and all care has been taken in compiling the
contents, neither the author nor the publisher and their
distributors can be held responsible for any loss, claim
or action that may arise from reliance on the information
contained in this book. As each person and situation
is unique, it is the responsibility of the reader to follow
government and travel authority guidelines regarding the
safety of destinations.

*To my parents, who always
understood the need to travel*

Contents

Preface

In 2019, I was invited to write a book with a simple premise: travel is easier and more accessible than ever before. Therefore, it needs to be taken *more* seriously. In the world of cheap flights and borderless continents, it was to be a rallying cry to eke more depth out of your experiences. To challenge your assumptions, break your psychological and information bubbles, and connect in a deeper way with the people and places you encountered. In the time of fear of 'the other', it was to be an invitation and practical guide to use travel as a means to connect on a more profound level with yourself, and with the world.

In creating this book, my goal was not to write from the perspective of someone 'better' at travel, smugly telling you how to do it. Instead, it would be a catalogue of the mistakes I made in my early, clumsy wanderings. Enhanced with the experiences of a diverse group of travellers of all ages and demographics who would provide interviews and insights, *Travel Your Way* would be a manifesto for the ambitious traveller. It wouldn't tell you to eat here or there, but it would contain advice for dodging tourist traps and finding yourself surrounded by locals, even in the world's top tourist hubs. It was to be the book I wish someone had written before I first set out on solo journeys at the age of seventeen.

Living in Budapest, I wrote the original draft of *Travel Your Way* among the cafés of the old town, seeking advice from many members of the Intrepid Times community of writers and wanderers, and receiving regular and valuable feedback from my publishers, Exisle. The manuscript was completed in late 2019, edited, proofread, typeset, designed and ready to hit the printers in early 2020.

And then the world I had set out to write about ceased to exist, at least for a time.

It was never my intention to keep travelling throughout the Covid-19 pandemic — it just worked out that way. My partner and I had moved out of our apartment in Budapest in February 2020, with the intention of travelling and working remotely for a month or two before settling elsewhere. So when we found ourselves locked down in Albania a couple of weeks later, we had all our possessions with us, and no home to return to. We stuck it out in Albania before drifting through Greece, Poland, Spain and, eventually, Tbilisi, Georgia, where I am writing from now.

While disconcerting at the start, in the end this was probably a twisted kind of good fortune: being unusually untethered made it natural to move about in the windows of openness that the world presented to us. We saw lockdown in three countries, but also experienced perhaps the warmest receptions of our travel careers — rarely have locals in well-trodden tourist cities on the Mediterranean been so thrilled to hear the clanking sounds of the English language on their streets.

Throughout this year, Intrepid Times co-editor Jennifer Roberts and I have mentored a small group of budding travel writers via the Intrepid Times Private Writers Club. Working with these mostly grounded travellers from the United States, Canada, the United Kingdom, India, New Zealand, Australia and beyond provided rich insight into how this year affected the mind of a traveller. Frustration and despair one day blended into hope and exhilaration the next. Each news bulletin brought with it fear or anger or exasperation. And throughout it all, the memory of past travel experiences grew stronger and more profound, as writers and travellers found ways to reconnect with the events, places and people that had helped form them.

For those of us who, like me, through luck or circumstance or recklessness, kept travelling throughout 2020 and the first half of 2021, the lessons we had acquired over previous years on the road often served us well in the new reality. The mechanics of travel may have changed, but the principles remain the same. Travel teaches you to study for the test, to treat each border guard like a snobbish school teacher, and present yourself to them like a preppy prefect. It teaches you to do your homework. To have the inane forms filled out. To know your rights. To look the part so you get waved through and spend as little time in administrative limbo as possible.

We might not always be able to control *where* we are allowed to go, but we can control how we interact with the people there. We cannot control the 24-hour news cycle, but we can control how we process information. What to believe and what to question. We cannot stop the world from trying to make us afraid to step out of our front doors. But we can keep the flame of wanderlust alive.

Not all the news, of course, is grim. In a world of remote workers, certain aspects of long-term travel are in fact more accessible now than ever before. You might find yourself less tied down to your office. Your bank or boss or mortgage broker may be more willing to engage with you via email. Those annual conferences might now be Zoom calls. Little by little, institutions seem to be starting to give you the freedom to conduct your affairs from wherever you choose to be.

There are other opportunities inherent in this moment. As fewer people travel, or more people stick to the familiar, those who do travel are more welcome than ever. As guide books become redundant and travel agencies crumble, people are being forced to engage their own creativity and grit and plot their own paths around the world, once again.

As barriers are erected against free travel, a part of us chafes at this imposition. Perhaps we had to experience the loss of travel to understand how much we really need it. And if so, the goal of this book has been moved closer, and not further away, by these otherwise ghastly events.

In rewriting this book to contend with the fast-changing new reality, I have tried to look beyond the day-to-day openings and closings, variants and vaccines, hopes and fears which still dominate the news cycle, and instead focus on what we humble travellers have the power to control.

This new, pandemic-proof edition of *Travel Your Way* is dedicated to its original purpose of helping you get more from your travel experiences. It is also galvanized with the understanding that your freedom to travel has never been more tenuous. In setting out concrete steps you can take to enhance your travels and get more out of those trips you are able to take, it is also my goal to fire up both your curiosity and courage as you venture out into this new world.

Travel, now more than ever, deserves to be done well.

Nathan James Thomas
Tbilisi, Georgia,
June 2021

Introduction:
Make travel great again

Travel is supposed to broaden the mind. To open us up to new cultures, expose us to new ways of thinking. And thanks to budget airlines, visa-free movement across much of the world and the proliferation of English as the *lingua franca*, travel is now easier and more popular than ever before. We should be living in an age of open-mindedness, tolerance and acceptance ...

But clearly something has gone very wrong indeed!

Travel *can* broaden your mind and challenge your assumptions, and has the potential to make you a better, smarter, wiser and more interesting person. But this doesn't just happen automatically. Anyone who has ever watched a group of boozed up twenty-somethings out on a stag night in Budapest, overheard an affluent westerner berate a waitress at a colonial hotel in South-East Asia, or listened to an eighteen-year-old on a Euro-bus tour brag about having 'done' France, will understand that travel alone does nothing. Even Trump travels.

It's not enough simply to be far from home in an exotic environment. Anyone can get on a plane. To really achieve the benefits that travel can promise, you need to go further. And I don't necessarily mean geographically further — there's very little of this planet that remains off the map. I mean psychologically further. To be willing to *listen* to the environment around you. To make an effort to learn a little of the history. Pick up a word or more in the local language. *Connect* on a more meaningful level with the people who live where you visit. Try to see the world through their eyes. And in doing so, deliberately, carefully and purposefully, you'll expose yourself to something new, and can allow it to change you.

In a world where folks on the political right would build walls to keep foreigners out of their country, and folks on the political left are happy to destroy someone's career over the crime of expressing an un-PC opinion on Twitter, what we need is to make *travel* great again. In this age of inequality and divide, taking travel *seriously* again, seeing it as a calling and a mission, can help heal and connect. True travel requires us to be humble. You cannot travel without first making yourself vulnerable. Putting yourself at the mercy of strangers, pressing pause on your own opinions, listening without prejudice, and discovering that, whether you're in China, Chile or Chisinau, us Homo sapiens really ain't that different from one another.

Sure, some long-term expats become *more* prejudiced, rather than less, due to their prolonged, often isolated exposure to a foreign culture. Plus there's the environmental downside — all those cheap flights are flooding the atmosphere with CO_2. The stampede of tourists to Venice is literally pushing the city under the water, and the rise of Airbnb is forcing locals out of their homes from Barcelona to Brooklyn. Any meaningful discussion of the merits of travel must take these aspects into account.

But travel undertaken from a sincere place of curiosity is the opposite of prejudice. It's the opposite of Trump's would-be wall. It's the opposite of the chorus of condemnation on social media. It's the opposite of the need to hurt, oppress, restrict or vanquish the 'other'. Travel is, instead, to *become* the 'other' in order to understand it.

This book explores ways we can get more out of travel, and in so doing become part of the solution to the divisiveness that is the bread and butter of politicians and pundits. It doesn't matter if you travel the world in order to eat *pho* in Vietnam, take selfies in the Sahara or write poetry about the craft beer scene in Poland. It doesn't matter if you travel alone, with your partner or in a group. It doesn't matter if you take a train, go by car or walk. It doesn't matter if you travel for three days or three years. This book isn't about finding the 'one true way' to travel, because that is a matter of personal choice and opportunity. Instead, we will look at how, with purpose, determination and a few tricks up your sleeve, you can find simple and profound ways to enhance your travel experiences. To arrive at a deeper understanding of each destination you visit. To sense more. Remember more. Learn more. And to connect, on a more personal and lasting level, with those you meet along the way.

The chapters in this book cover different ways for you to enhance the travel experience. Some explore attitudes to specific destinations and explain how putting an often-maligned place in context can help replace fear with curiosity and compassion. Others explore specific approaches, techniques and strategies for making your travel experiences count. You'll learn how to gradually rid yourself of the barriers and prejudices you may have without knowing it. You'll learn how to get to know locals on a meaningful level, even if you don't speak a lick of the local language. You'll learn how to overcome feelings of embarrassment and embrace the awkwardness that is at the core of the travel experience. You'll see how true travel can be achieved even in your hometown, and how meaningful travel is about the attitude you bring as much as it is about where you go.

The ideas and principles in this book are about travel, but they shouldn't be limited by geography. That is, after all, kinda the point: whether you're a New Zealander discovering Italy, an Italian discovering Australia, an Indonesian discovering the United States or an American discovering Great Britain, the ideas, attitudes and principles in this book apply to *you*, no matter where you're from or where you go. By the end of this book, you'll understand that real travel is just as achievable in your home town as it is on the other side of the world.

It is not my goal in these pages to rehash the clichéd distinction between being 'a traveller versus tourist', or to explain how one style of travel is in some way superior to others. Instead, we're going to explore some ideas that can help you to get more from the time you spend as a stranger in strange lands. By doing so, in this age of walls, we can each at our best become a bridge between cultures, between worlds, and between points of view. So, welcome aboard — let's get this journey started.

1.
Putting a place into context

The walls were of grey, exposed concrete. Raw meat hung from hooks suspended on the ceiling. Spectral figures loomed out of the dark corners, clutching pint-sized bottles of vodka. Children gazed out from the arms of destitute mothers, imploring the passing commuters to hand over a coin. At least, that's how I remember my first experience of Poland — the country of many of my ancestors, and the land that would later become my adopted home for several years. Katowice train station, 2010.

Since then, the station has been thoroughly renovated. The old Katowice train station was probably never as bad as it appears in my memory, though my Polish friends all have horror stories of their own — it was notorious. Arriving there alone, off a train from Berlin, at the tender age of eighteen, I was utterly terrified.

Katowice took me to Krakow, where my sense of anxious excitement at being somewhere so *different* was quickly punctured by the sheer number of other tourists I found in the city. Poland in my mind was the exotic far-east of Europe. I had fantasies of myself as the wide-eyed wanderer on the road to discover strange and forgotten lands and bring back to New Zealand an enviable swag of impressive travel stories. Hostels full of fellow Aussies and Kiwis happily drinking beer did not fit with this vision. I needed to push it further.

Something glimmered on the hostel wall. The room filled with an inspirational soundtrack — *Chariots of Fire*, perhaps — and some enchantment seemed to lift one brochure off the shelf beside the reception and levitate it gently into my hands. The thin paper brochure advertised a new youth hostel in a city called Lviv. Ukraine was just an overnight train journey away. How could I resist?

My knowledge of Ukraine was ... well, that it was a country. It pretty much stopped there. I'd never heard mention of it in conversation, or knowingly met anyone who had been there. Today, of course, the country is frequently in the news, its citizens are an established presence in the European Union, and its battle against the Russians is common dinner party conversation. And maybe it was also a hot topic in 2010. But for me — fresh out of high school and on the other side of the world — it was like an invitation to Narnia. No. It was stranger than that. People I knew had heard of Narnia. This was an opportunity truly to go somewhere that no one in my circle had ever gone before.

I set out that night.

Across the border and into the unknown

Several loud thuds on the door of the sleeper carriage jolted me from an anxious sleep. Guards walked in and checked our passports. They handed mine back, smiled and left. I was through. Or so I thought. A few minutes later the whole train shook as if taken by some hideous seizure. We stopped, lurched forward, and then stopped again. More thuds on the door. A sleepy passenger clumsily reached over and unlocked the door. We had thought we were through. Turns out, the first check was a farewell from Poland. Next stop, Ukraine.

Three guards, all in my memory enormous, bald and armed with Kalashnikovs (this is surely an embellishment of my memory, but they were without a doubt equipped for battle) stormed into the carriage and started screaming about passports. Once again we handed them over. But this time, rather than the cursory glance afforded at the Polish end, they took them all and left. Terrified, we waited in silence as the minutes rolled by.

Finally, we made it. The old train station shone like a mosque in the pre-dawn darkness. The cars were old. The writing was in Cyrillic — my first time in a country using this ancient-looking script. There were no other foreigners at the station. I heard not a whisper of English, no suggestion of a youth hostel pub crawl, no invitation to enjoy an 'authentic local meal' at 10 per cent off. I was in heaven.

For the next few days, I wandered around the city in a state of utter bliss. I was the only one in the dorm at the 'hostel' I'd seen advertised in Krakow. The owner quickly summoned a friend of his to show me around. A young man with a thin, black beard, the friend made his money in the antique coin trade — his practice was to smuggle coins across the border into Poland, where he sold them at high margins and avoided tax. Together, we hit the eccentric bars and smoky cafés of this city that, to me at that time, was on the edge of the world.

This was the dream of travel realized. This, I thought, made me better, braver, cooler, wiser than my fellow Aussies and Kiwis on their pub crawls in Krakow. I am truly embarrassed to admit it, but some part of me really thought that way. But having spoken in later years to other travellers, I have come to understand that this self-mythologizing travel kick was nothing special. It's something most of us born with a bit too much wanderlust constantly strive for. And even today, some part of me is searching for another taste.

It's a wonderful feeling. But like a drug, it gets harder and harder to recreate. I've been back to Lviv on a couple of occasions since that time. The city is now a haven for Polish tourists. It's friendly, familiar and accessible. That old excitement now exists as an ever-weakening nostalgia. I've searched for it elsewhere. In Iran, Turkmenistan, Tibet, along the banks of the

Danube and under the palm fronds of the Mekong. It still comes. But you have to work ever harder for it.

Ignorance can fuel curiosity; it can also impede it

Knowing nothing of a place is a quick and easy path to fascination: the unknown is the exotic. Being there is thrilling. And it is valuable: you colour in the lines on the map with your own experience and meet locals who give you an emotive, human sense of the place where you are. This is real and valuable travel. But it is a diminishing high. As the mere thrill of being far away from home fades (and it does), you need to search for more ways to connect with where you are, more pathways into the core of the travel experience.

Putting a place into its context increases the thrill *and* makes it last longer. It is the beginning of the process of shifting your focus away from *you* and into the place where you are. It infuses each experience with more meaning, because direct observation in isolation can only provide so much. It is the difference between the fumbling excitement of a one-night stand and the profound surrender of falling in love.

On that first trip to Ukraine, I must confess that I knew nothing of where I was. I did not know that Lviv was for many years part of Poland, that the citizens from Lviv had been shipped to Wroclaw, which was taken from Germany to become part of Poland as the borders of Poland shifted west in the wake of World War II. This might have infused some meaning into the similarities I observed in structure, building and layout from Krakow. I didn't know about the forces that were then, as now, threatening to pull the relatively new country apart. The powers of Russia and the European Union vying for influence, with Lviv squarely on the side of the latter.

Even the thinnest knowledge of historical context may have infused my observations and conversations with greater depth. I may have noticed things then — the protesters, the tensions, the look and attitude of the locals — that would today greater inform my understanding of one of the most contentious conflicts in Europe. So while my time there was a thrill, it was also to some degree a missed opportunity.

The trouble with being a happy-go-lucky blank slate is that you don't know what questions to ask, and you have no framework in which to structure the incredible amount of data that is presented to you. A gouge in the wall of the General Post Office in Dublin is just an error in construction unless you know a thing or two about the Easter Rising. Uluru is just a big rock until you know it has been a sacred site of Australia's Indigenous people for tens of thousands of years. The Gran Café de Paris in Tangier is charming but a bit bleak unless your imagination can colour in the empty seats with the interzone writers and misfits who fill the pages of William S. Burroughs' novels.

Filling in the blanks requires a greater commitment to the experience of travel. Hopping on that first plane ride after over a year of isolation is a thrill that you deserve to enjoy, but that rush won't last on its own. It's not sufficient simply to be somewhere if the real depth, meaning and story behind the place where you are is lost on you. This is fine if you're taking a holiday or you're on a business trip. But if your goal is to Travel with a capital T, and you want to live up to the full promise of Travel, then it's essential to commit some time and energy into investing in the experience and building the scaffolding of knowledge that will allow you to gain greater understanding and insight from your trip.

So there's nothing for it but to postpone those Ryanair flights and block out the next few evenings for candlelit study of history, culture and linguistics at the local library, right? Sure, there are definitely some nights that I spent in backpacker bars that could 'perhaps' have been better spent in a library, but that ain't the message of this book. Here's the thing: back when travel was a more challenging and expensive undertaking, this advance research was a part of the process. Your survival may well have depended on it. Now, you just need to book a flight on Skyscanner and call an Uber to the airport.

We made travel easier than ever before, and we took it for granted. And although we may now yearn for that ease and simplicity that seemed to vanish in March 2020, we can also use this moment to pivot our mindset. To aim to get the most out of every destination, every experience. To cherish and treasure the wonderful joy and opportunity that travel truly is. To drink deeper from the cup.

And even when (and surely it is a when) travel is once again as easy and ever-present as it was in 2019, we can continue to search not just for thrills or passport stamps, but for understanding. This simple change in attitude is really all it takes to get more from travel, whether you're free to backpack around the world, or simply dashing across whatever border happens to be open this week.

And if you're pressed for time, in a rush, or simply have other things on your mind before you jet off on your trip, you can begin this process of learning to understand the places you visit, as I often do, when you're already there ...

> *Solo travel enables me to discover how wildly capable I am. Flung far from the familiar, it pushes me to jump into the unknown, from swimming with great whites and encountering a hyena on my veranda in South Africa to nearly being tossed off a camel and stuck on an ATV in the Arabian desert. The fact I continue to come out unscathed just shows me how remarkable and wonderful this world of ours truly is, for I may travel solo, but I'm never truly alone!*
>
> Sarah K, Canada

Building the scaffolding of knowledge

The path to understanding a place isn't about memorization of facts. Anyone who has suffered through a geography class in school will get that knowing the fact that the capital city of Uganda is Kampala, its population is 42 million and the president's name is Museveni tells a traveller basically nothing of real use. Understanding a place, like understanding a person, is a matter of story.

Knowing a person's backstory makes them interesting, it puts their activities in context, and it helps you understand their frame of mind and the motivation behind their behaviour. I've found it helpful in my own tentative understanding of the world to practise looking at destinations in the same way. As we will explore later, a 'country' is a very imperfect and often artificial label slapped on to a complicated mesh of competing identities. But to know a place, any place, is to know — to whatever degree a thoughtful yet transient visitor can — the lived experience of the locals.

To understand a place, even to the slightest degree, is to aspire to know the events and ideas that shape the people who live there. It is to be able to walk, if not a mile then at least a few paces, in their shoes. Of course, unless you're willing and able to relocate to a place for several years, intimately learn the language and mingle with people from all strata of society, you will never perfectly fulfil this objective. Who among us even fully understands where we come from? Like your own homeland, the places you visit will be complex, made up of not one but several identities, all of which are intimately rich, complex and worthy of appreciation rather than mere simplification.

But we have only one lifetime. It is impossible for even the keenest, most committed scholar to gain a worthy appreciation of every place and culture on this Earth. Not to mention that most of us are part-time amateur travellers, balancing our wanderlust and curiosity about the world with our own ambitions and challenges as citizens of our distinct homes. With family. With career. With this thing we call life. We have to do our best with the limited time and resources that we have. I have found that the best way to gain some sense of understanding and deeper appreciation of a place I am travelling to is to work to establish a sense of that place's geographical 'centre of gravity'.

The geographical 'centre of gravity'

A place's centre of gravity is a simplification of how the locals define themselves in relation to both their own history and the outside world. The thinking is that in order to deeply look into a place is to be able to look out from it. To be able to, in whatever small degree, observe the world from the perspective of the denizens of that place.

Immediately, of course, we encounter a problem. The idea of 'the locals' of a place as some sort of distinct, homogenous entity is an oversimplification. What does a 'local' of the United States mean? Does it mean an African American from Alabama, a native American family in Idaho, a second-generation Polish family in Chicago, a Hispanic family in Texas, or a descendant of the *Mayflower* settlers in South Carolina? All of these people are American. Yet all would have radically diverse identities and define themselves through different events. However, there would be some commonalities between all of them, something that would define them as 'American'. They would all, most likely, experience a broadly similar visceral reaction to the horrors of 9/11. Mexico would likely occupy a greater percentage of their mental space than New Zealand. Canada, equally close geographically, would unlikely be thought of as a threat. Chances are, they would see democracy as a human right, and the idea of human rights would be important to them.

Let's try the same thought experiment with China.

China, a country of around 1.4 billion people, is composed of 56 different officially recognized ethnicities; however, the vast majority of the population is Han Chinese in ethnicity. Most of China is officially atheist in religion; however — particularly in smaller cities (for small, read fewer than 2 million residents) and rural areas — superstitions that blend practices of Buddhism and Taoism are widely held. The life experience, identity and aspirations of a Han Chinese businessman in Shanghai would be radically different to that of a Uyghur Muslim in Xinjiang. However, for most of China there is a sense that this is now China's century, that their country is on the right path and that, though they may have made mistakes in the past, their government is steering the ship in the right direction. They would likely see the United States as powerful yet slightly absurd, understand Japan through the extremely negative lens of the Nanjing Massacre and the more positive lens of their desirable products and technology, perhaps think New Zealand is a paradise where milk comes from, and possibly view Taiwan as a rebellious and illegitimate breakaway province whose reunion with the mainland is inevitable. As for human rights and democracy, as many people in China have told me, they are 'foreign ideas' that 'would not work for us'.

Many people in both these countries would reject being pigeonholed into this mass perspective, and they would be right to do so. The point is not that this oversimplification of a country's centre of gravity precludes you from learning more; it empowers you to learn more. It lets you know the right kind of questions to ask, so that your experience as a traveller can add texture, nuance and colour to your understanding of this basic black and white picture. For example, many western readers of this book would be surprised, as I was the first few times I heard it, to learn that there are those in China who consider 'human rights' to be a foreign and alien concept. Hold on. How can anyone disagree with the idea of human rights! Not everyone in China does, of course, but isn't it an exciting thing to be challenged on one of the fundamental assumptions that defines your understanding of the world — an assumption you have probably taken for granted and never even properly examined until now? (And does it not make you wonder just what ideas do other countries hold to be self-evident and irrefutable, that you would consider completely alien?)

Locating the centre of gravity

You probably have a good sense of what the United States 'is'. No matter where you're from in the world, the United States is coming at you in the form of music, movies and media. Everyone follows American politics. Forming some picture of America's geographical centre of gravity is natural. Of course, upon travelling there, some of your assumptions will be challenged and you'll most likely be forced to change some of your ideas. And if you are an American reading this, you will observe this same phenomenon in reverse: those you meet overseas will likely have set prejudices about how you as an American ought to 'be'. And you, as the unique and complex person that you are, will challenge this.

Both ways round, these assumptions are kind of a good thing, if used properly. When applied with curiosity, prior knowledge like this provides some sense of what questions to ask and what vocabulary to use. 'Hey, China has rapidly changed from a rural to an urban country — how has this change affected your family?' 'Hey, you're an American. How the bloody hell did Trump happen?' At least the conversation is started! This is a new way of thinking of 'prejudice', as a sense of questions to ask and assumptions to be challenged, rather than as a closed door.

While it's nothing to celebrate, the shared experience of having gone through the Covid-19 pandemic is itself an opportunity to break down barriers. 'What was Covid like here?' You may well be about to get hit with a torrent of conspiracy or vitriol, but even that itself is informative. Hold your own opinions softly, and invite others to open up. This may just break the ice and reveal to both parties how much you really do have in common.

But what of other countries for which you have not been granted so much prior exposure? What of the countries about which you have more ignorance than prejudice? After all, there are 193 countries out there according to the United Nations — no one simply wakes up with a preconceived idea of all of them. So how do you handle arriving somewhere as a blank slate? How ambitious you want to be with this depends on the level of commitment you have to the place you are going to. Of course, if you're moving to a place, the level of research you should be doing is quite different than if you're simply passing through for a couple of days! But even if you're just stopping by, knowing the answers to a few simple questions will help you contextualize what you observe and gain more from the travel experience, however fleeting.

Searching questions for nosy travellers

Here are some general questions that can be applied to more or less anywhere to help you start the conversation, gain more information and structure what you learn in a way that will help you better understand where you are.

- How long has this country been independent? This will help you understand the greater cultural and historical forces that have shaped this place, as well as any nationalist ideas that the locals you meet may hold dear.

- Who, if anyone, was its previous colonial master, or what empire was it a part of, and how do people feel about it? Do they feel as if they are the recently freed (or still imprisoned) victims or oppressors, the proud conquerors of great swathes of the world, or none of these?

- What is the average income, and how does it compare to that of the country's neighbours? The comparison is key — Cuba may seem poor compared to Australia, but it's richer in practice than much of the world. Poland is a rich and successful country by global standards, but the fact that it is still poorer than neighbouring Germany greatly influences the country's politics.

- How much freedom do people have to vote, protest or travel? The philosopher Karl Popper saw the world comprised of two types of countries: those where there is 'one truth' that everyone must bow down to, and those where people are free to choose, debate and argue about their own version of the truth.

- Who is the country's leader? How and for how long has this person been in power? Asking people their opinion of the national leader, and understanding from where this leader gets their power, can be a wonderful — if risky — conversation starter! Just be careful of this one when travelling in less free countries where people may be afraid to express political opinions.

- What are the main ethnic groups who live in the country? This leads into a subject we will discuss in more detail later, about sources of identity within a 'country' and how the concept of a 'country' itself often fails to take into account the distinctions of the people who live within its borders.

- What are the main languages spoken? If it's more than one, then who speaks minority languages, and why? How prolific is English? If most young people speak English but most older folk don't, that may tell you something about how a place has experienced rapid change in relation to the outside world.

Five minutes on Wikipedia could begin to furnish basic answers to most of the above. Knowing the simple answers to points alone may not mean much in isolation; however, they provide the tools ('signposts' if you will) that will allow you to learn more when you're actually there.

When I travelled briefly through Kazakhstan a couple of years ago, vague answers to the above questions were basically all that I knew. However, even this level of knowledge was enough for me to gain a sense of meaning when people-watching, and when I was lucky enough to meet an English-speaking Kazakh of Korean origin while on an overnight train journey from Almaty to Astana, this little 'Wiki-knowledge' allowed me to ask the right kind of questions that led to a richly informative conversation.

With this basic starting point alone, you'll put yourself a cut above most and, most importantly, create a basic framework which will allow you to ask smarter questions, put what you observe in context, and gradually add more layers of meaning to this initial overview. If a country has multiple official languages, knowing which language a person favours — for example, Russian or Kazahk in Kazakhstan — can tell you a lot about how someone sees themselves. And while Kazakhstan may be 'poor' compared to, say, Germany or Switzerland, it is vastly richer than its neighbours, a fact that locals I spoke to were proud of. In conversation or in your research, you can then choose to take a deep dive into any one of these areas, while being sensitive to the fact that people from oppressive countries are often cautious about voicing a direct opinion on political topics.

The 'events' lens

If you have more time to get to know a particular place, looking at its recent history through the lens of major events can provide a fantastic opportunity for doing a 'deep dive'. In the national psyche, most places have a few events by which they define themselves. In the United States, this could be, for example, points of conflict and struggles for freedom: the War of Independence, the Civil War and 9/11. In Hungary, as with many of the much-bullied Eastern Bloc nations, defining events are often revolutions: 1848 against the Habsburgs, 1956 against the Russians. Choosing to select one of these events and read and learn about it in detail can give you a wonderful window into how history works in a particular place, a microcosm of the forces, attitudes and influences at play.

Events never exist in isolation, and doing a deep dive into one story will inevitably introduce you to other ideas and forces at play. When I moved to Budapest midway through 2019, I picked up a book — more or less at random — on the 1956 uprising. This greatly helped enhance my understanding of the city: I would recognize landmarks in my everyday wanderings that were a key part of the events, and understand the story behind the bullet holes on the walls of otherwise ordinary buildings in the district where I lived. This naturally led me to discover more about how Hungary developed post-1956, during the reign of Janos Kadar, and perhaps come to understand at least something about the country's complex relationship with Russia and with its own leaders. The '56 revolution echoed the 1848 revolution, about which I'd mostly been ignorant. I lived near Petőfi bridge, and so came to understand a little of the story of the firebrand poet who this landmark was named after. The 1848 revolution in Hungary was, I discovered, just one of a series of revolutions that took place that year across Europe. And so on down the rabbit hole …

This prior knowledge will not make you a cocky know-it-all. Used properly, it will humble you and give you a sense of what to search for, aiding your appreciation of where you are. The American travel writer and professor at Drury University Elizabeth Gackstetter Nichols gives us a sense of this in a story published on Intrepid Times, the travel writing magazine that I edit, about her search for the grave of the legendary Native American linguist, guide, mother and adventurer, Sacajawea. Wandering alongside her daughter in the 'sea of dust' that is the Wind River reservation in Wyoming, Elizabeth reflects:

> I was hoping to pay my respects to a woman who my daughter and I viewed as courageous and skilled. I wanted to recognize her (often ignored) contributions to US history. But what did I know about Shoshone history and reality? Very little. I feared that this made me an intruder, an inappropriate glommer-on.

And yet it was this humble curiosity that led this writer on in the first place. As Elizabeth reflects at the end of her article:

> I still didn't know the answers to many of my questions, and I left resolved to learn more about the Shoshone. I also left feeling that I had visited the grave of a woman who continues to serve as a symbol of strength and intelligence, a symbol of skill and perseverance — not only to my own child, but to the children of a tribe proud to number her among their heroes.

A more ignorant traveller would not be aware of their own ignorance. They would not be aware of the significance of where they are, and would not know how to fill the gaps in their own knowledge, as they would not even know that those gaps existed.

Using the prior knowledge you have gained in your reading and research, also pry gently yet persistently when given the opportunity to speak with locals. When doing so, follow their lead — the events that are important to them may not be the ones you expect. In the West, when we think of China we may think of the Cultural Revolution and the Tiananmen Square massacre. But for your average local in mainland China (the product, of course, of a politically sensitive education system) the Nanjing Massacre of 1937–8 and reunification with Hong Kong in 1997 are events likely to occupy more mental space and lead to more interesting conversations. Be willing to follow whatever rabbit holes you happen to stumble upon.

TRAVEL PRACTICE
What is your 'centre of gravity'?

Have a look back at the centres of gravity we explored for the United States and China (see pp. 22 and 23). And, using your current knowledge, perspective and understanding, try to write, or at least think out, the same for your own country. Think about the ways your own perspective of the world is shaped by where you come from. And think about how you would go about explaining this to an interested yet rushed traveller. Where should they start? What books should they read, and what music should they listen to? Who should they speak to, and what questions should they ask? Then apply the same questions to another country that is not your home, but is somewhere you are familiar with — perhaps somewhere nearby that you have visited frequently. Next, challenge yourself to do the same, within the limits of your time and resources, with the next place you plan to visit.

To help you get the ball rolling, let's start with a challenge, and try to create a sense of historical understanding from the three countries that are arguably the most 'foreign' to our western perspectives: the so-called Axis of Evil.

The Axis of Evil

We'd crossed the border from Azerbaijan that morning and were fighting our way through the dense Tehran traffic when it happened for the first time. A taxi driver had stopped next to us at a red light. His face was deeply lined, his mouth obscured by a huge white moustache. He noticed the four foreigners cramped into the bizarre, clapped-out car parked next to him. He wound down his window and stared for a moment.

We held our breath — worried that we were about to get reported, shouted at, heckled for being the foreign heathens we were. The taxi driver reached forward and undid the ornate wooden chain that was dangling from his rearview mirror. He motioned to us to wind down our window, and we did so. Leaning out of the window, he handed the ornament through to us and smiled. 'Welcome to Iran,' he said, before the light went green and he shunted forward and was soon lost in the traffic.

The image in my head of Iran was that of bearded *mullahs*, women obscured by burqas, anti-American propaganda, religiosity and loathing of the West. And much of that is there, if you look for it. But we also found the most welcoming people you are likely to ever encounter in your travels. Open, curious and friendly, they were delighted that we westerners had come to see Iran for ourselves.

But few visitors make the trip. The Axis of Evil brand was never good for tourism.

Iran, North Korea, Iraq

On 29 January 2002, barely five months after the 9/11 attacks, George W. Bush stood before Congress and announced that North Korea, Iran and Iraq would be referred to as the Axis of Evil. A clever use of words concocted by the Bush speechwriter David Frum, the phrase evoked the Axis powers of World War II and provided a clear narrative that pitted the United States against the 'evil' sponsors of terrorism. While North Korea isn't known for orchestrating acts of terrorism, its constant attempts to arm itself with 'weapons of mass destruction' (i.e. the nuclear bomb) earned the small Asian nation a spot in the naughty corner alongside the Islamist nations of Iraq and Iran.

While George Bush's comments took aim at the regimes that governed these nations and portrayed their citizens as beat-down and oppressed, the labelling — among the harshest use of words to describe a foreign power by a president during a formal address — was decisive. The political decisions immediately coloured the minds and perceptions of everyday Americans. And, due to the massive cultural influence of the United States, other western citizens as well. These countries were *evil*. The president of the United States just said so.

The Axis of Evil label even influenced the country's perceptions of themselves, and it certainly did nothing to dissuade any of these countries from their 'evil' ways. Since being labelled as 'evil' by the so-called 'big Satan' — the United States — nationalist and far-right rhetoric in Iran increased at the expense of more moderate political dialogue.[1] US relations with North Korea had improved during the Clinton era. Missile tests had stopped, and there was even talk of a possible visit by Bill himself, but the moment the speech was delivered, relations began a rapid deterioration which — Trump and Kim Jong-un's summit theatrics notwithstanding — has continued to this day.[2] As a result of this labelling, the gulf between the United States and these countries widened. Fear, anger and blame ensured there would be no attempt at mutual understanding. There would be no *travel*.

Labels such as this, depictions in the media and the pounding rhetoric of hawkish politicians, causes us to form in our minds an image of such countries and their citizens. Most of us are probably not even conscious of how we came to believe what we do. Many of us have our reservations about the Bush doctrine. We may criticize the politics that led to the 'Axis' labelling. Yet some part of our minds will still be tempted to regard these countries as evil. The bubble we live in is a hard one to burst. Like the fish who has never heard of water, we often are unaware of the sea of assumptions in which we are swimming. But how much about these so-called 'evil' countries do we actually know?

It's true that these and many other countries are and have been ruled — largely against the will of their own citizens — by extremists and tyrants.[3] The Axis of Evil countries have long histories, but the origin story of their current tyrannies can each be traced to the 20th century, and to the meddling of the very western powers who have been so eager to name-call in more recent years.

Revolution '79

In 1953, Iran's Prime Minister Mohammad Mosaddeq nationalized the British-owned Anglo-Iranian Oil (which later became BP). Calling on their buddies in the CIA for help, the British government's military intelligence wing, MI6, organized a coup that overthrew Mosaddeq's government. Working together, the British and Eisenhower governments isolated Mosaddeq from the international community, economically suffocated his government, and used so-called 'covert political action' to mobilize powerful business and political figures against Mosaddeq.[4]

Mohammad Reza Pahlavi, the monarch Britain helped install back in 1941 but who was pushed out by Mosaddeq, was once again in power thanks to Britain. But the Shah quickly found himself falling out of favour with the political and business elites. His so-called

'White Revolution' rapidly modernized Iran, stripping powerful landowners of much of their property and causing a rapid shift in cultural values that even secular elites were uncomfortable with.

Like many powerful movements, Iran's 1979 revolution began with the students. In January 1978, young religious school students protested in the streets of Tehran, many declaring their support for Ayatollah Ruhollah Khomeini, a philosopher and religious hardliner who had been exiled from Iran for speaking out against the Shah. Called to the streets by the rallying cry *'Allāhu akbar'* — God is great — more and more young people joined the protests.

Reza Shah's clumsy and violent attempts to suppress the political movement against him only escalated the situation. Soon, he had lost support among the businessmen and elites upon which his regime depended. In time, 1 million people had taken to the streets in Tehran, voicing support for Khomeini. He returned to Iran in February 1979. Ten days later, the Shah had fallen.

Brought to power on a populist wave that opposed modernization and westernization of Iran, it was expected that Ayatollah Ruhollah Khomeini would usher in an era of religious orthodoxy, alongside all the associated oppression. In many ways, this was true. Khomeini's philosophy was *'velayat-e faqih'* — guardian by the leading Islamic jurists. Sharia law was established, meaning governance in accordance with

Islamic law. Alcoholic drinks were banned, and Islamic dress was imposed for both men and women. Efforts were made to suppress opposition. Ironically, the CIA smuggled the Islamic government a list of KGB agents in Iran, who were then rounded up and executed.

Confusingly, Khomeini was succeeded in 1989 by Sayyid Ali Hosseini Khamenei, whose name is inconsiderately similar to that of his predecessor. Khamenei continued the 1979 Islamic theme of his predecessor. Women's rights stalled: the 2017 'Global Gender Gap Report' published by the World Economic Forum — which rates countries' treatment of women based on four categories (economic participation and opportunity, educational attainment, health and survival, and political empowerment) — places Iran at 140 out of 144 countries.[5] Women have to wear hijab in public. Shielded from the sun even in the middle of summer, vitamin D deficiency is a major health concern for Iranian women.[6]

But in some areas, Iran has improved. Rates of illiteracy fell from 52.5 per cent in 1976 to 6.2 per cent in 2002, according to the United Nations Population Fund.[7] Driven by the pragmatism of the legislators who have governed under the two Ayatollahs post-revolution, the poverty rate fell from 25 per cent in the 1970s to less than 10 per cent in 2014.[8] While the power of the clerics, overseen by the Ayatollah, is supreme, citizens elect their president. The latest election, between Hassan Rouhani, largely seen as relatively left leaning, and Ebrahim Raisi, saw the candidates competing not for kills on the street, but for likes on Instagram. Rouhani won.

Iraq: Desert Fox

As with Iran, the fingerprints of the British Empire are all over the crime scene that is today's Bagdad. As the Ottoman Empire collapsed during World War I, Britain found itself in control of much of what is now Iraq. These lands in the historical region of Mesopotamia — the proud area that had seen the development of the wheel, astrology and mathematics, for a start — were quickly handed to Britain by the League of Nations, the precursor to the UN.

Unsurprisingly, the Iraqis weren't all that chuffed about suddenly coming under the thumb of the British, and in 1920 they rebelled. In the quaint words of the BBC, the revolution was controlled 'only with great difficulty and by methods that do not bear close scrutiny'.[9] Poison gas and relentless air assaults were likely involved. The result was a fabricated 'Iraq' that unified many different tribes, religions and ethnic groups under the fragile dominion of a now much-resented foreign power.

In 1932 Iraq was granted independence, but British influence lingered. Independent Iraq had to accept the borders clumsily drawn by the British. Tellingly, these borders cut Iraq off from Kuwait, an act that caused deep resentment among many Iraqis. Iraq's invasion of Kuwait on 2 August 1990 under the command of Saddam Hussein triggered the First Gulf War and led directly to the Axis of Evil speech, operation Desert Fox, and the second American invasion.

Britain continued to exert influence in Iraq — mostly in the interests of protecting their oil fields, you will be shocked to hear — until a nationalist coup in 1958 brought Iraqi hardliners to power. In 1961 Iraq's leader, General Kassem, attempted to seize Kuwait, but this was blocked by British troops. Kassem was kicked out in yet another coup — this one most likely organized by the CIA, cocky after their 'success' in Tehran — creating a vacuum that eventually allowed Saddam Hussein to seize power in 1979, the year of the Iranian revolution.

Hussein sought to take advantage of the chaos over the border and launched an invasion of the rival country, using weapons and intelligence supplied by the United States and Britain, who were more than happy to see Hussein take on revolutionary Iran.[10] The fighting took almost ten years, saw the use of thousands of child soldiers, and resulted in a stalemate. So Hussein turned his mind back to Kuwait.

As the British had done in 1961, the Americans successfully defended Kuwait from Iraq. Doing so required the largest military force seen since World War II. Together, the coalition of Kuwait, the United States, the United Kingdom, Saudi Arabia, Egypt and France spent six months and US$60 billion keeping Iraq at bay. Following the war, the United Nations located and destroyed Iraq's chemical weapons supplies.

After their defeat in the Gulf War, Iraq had promised to cease their chemical, biological and nuclear weapons program. George Bush and Tony Blair claimed they had intelligence proving that Iraq had broken this promise and acquired weapons of mass destruction. When Hussein refused to fully comply with UN inspectors, Bush launched an invasion in 2003 that toppled Saddam Hussein. No WMDs were later found.

Given a history that makes even Iran's story seem like a children's picture book, it is unsurprising that today's Iraq is no safe haven. In the rubble of the second Iraq war, the terrorist cell ISIS, which sprang from Al Qaeda, was able to take advantage of the chaos and seize control of Iraqi territory, declaring themselves a caliphate. Though their territory has been reclaimed, they are still a potent force for destruction. Iraq continues to be wracked by sectarian conflict, with hundreds of thousands of citizens displaced from their homes. But there is hope. The United States Institute of Peace claims that many internally displaced persons are now able to return home.[11]

Despite the high risk caused by the ongoing internal conflicts, civilian life in Iraq does continue. Around 40 per cent of Iraqis were born after the US invasion.[12] They didn't grow up listening to the anti-US rhetoric. Instead, they grew up watching YouTube. They don't want to join ISIS — they want to be Steve Jobs. Baghdad is even home to its own start-up incubator, The Station. You can find it on Twitter, Facebook and Google Reviews, where it holds a 4.5/5 rating. One recent review describes it as a 'young, hipster place with great coffee'.

The hermit kingdom

Everyone knows North Korea is *crazy*. We know there was a war in the 1950s with China supporting the north and the United States supporting the south. We know there was a line drawn between the two new states. The north became communist and went mad. The south became capitalist and got rich. This narrative has fairytale-level simplicity. It makes us westerners feel good about being on the 'winning' capitalist side. But of course, there's more to this story, too.

Like the Middle East, the Korean Peninsula has been a battleground where different international powers have vied for influence. Like Iran and Iraq, when North Korea finally gained independence it did so with a vengeance, and an autocratic regime took over with a nationalist ideology characterized by a suspicion of foreigners. While Iran and Iraq dedicated themselves to versions of Islam, North Korea took up the 'religion' of communism, and the role of the villain in this story is played not by Britain, but by Japan.

In 1905, Japan fought a war against Russia over the control of strategically valuable ports in the Korean Peninsula. Japan won, embarrassing Russia and emboldening Japan's imperial ambitions. Korea became a protectorate of Japan in 1905, and then following backhanded negotiations in Switzerland, was wholly annexed by Japan in 1910.

Japan's rule over Korea was brutal. Korean language and history were banned in schools, and people were forced to learn and speak Japanese. Dissidents were arrested and often killed. Absolute loyalty to the Japanese emperor was demanded of all citizens. In this, we see the seeds of North Korean totalitarianism, which would become an unholy blend of the fervours of Japanese imperialism and the madness of communism. Korea continued to mount resistance movements and, while at war against China in 1941, Japan had finally had enough. They attempted to extinguish the entire Korean culture, banning everything that was not strictly Japanese.

When the Japanese empire collapsed in August 1945, the victorious communist and capitalist powers found themselves vying to pick up the pieces and shape the newly independent Korea that had been promised to the Korean government in exile. Of course, the Soviet Union, in alliance with China, and the United Kingdom could not see eye to eye, so a compromise was reached, and Korea was divided at the 38th parallel. Syngman Rhee was given power in the south, accountable to the Americans. Kim Il-sung was given power in the north, accountable to the Soviets.

Behind the 38th parallel, both sides had been building up their armies. The Chinese were aiding the north, who were winning the arms race. On 25 June 1950, the north invaded. The south was quickly in rapid retreat. The United States sent soldiers to push the north back. Then China sent more soldiers — up to 1.2 million in 1951 — to bolster the north. The two sides were eventually locked in a stalemate on the 38th parallel — exactly where they had been at the end of World War II. Four million people had died.

The rest of the story is familiar. Ruled by the Kim dynasty, North Korea maintained close relations with the Soviet Union, purged enemies, harassed the south, collectivized farms and marshalled workers into state-run factories. All the while, the ruling family lived a lavish lifestyle by treating state funds as their own personal bank account. For a while, the state functioned, more or less. Propped up by the Soviet Union, North Korea under Kim Il-sung was not quite as impoverished as it is today. But the collapse of the Soviet Union, followed by a series of natural disasters in the 1990s, sent North Korea into famine. Isolated from the world, desperate and starving, they needed a way to maintain the survival of the regime and keep foreign 'enemies' at bay. The solution: nuclear weapons.

Since they were accused of violating the Nuclear Non-Proliferation Treaty in 1993, North Korea has been fanatically devoted to acquiring a nuclear arsenal that would strike terror into the heart of their enemies. It is their single-minded pursuit of so-called 'weapons of mass destruction' that earned North Korea a spot in Bush's Axis of Evil list in 2002. When the 27-year-old Kim Jong-un took over from his father Kim Jong-il in 2011, he continued to pursue nuclear weapons and would deftly use his nuclear program as a way to win attention and extract concessions from the United States. While some hope the north will eventually give up their pursuit of 'nukes', the survival of the Kim dynasty depends on the power that fear of this arsenal gives them.

South Korea, now a thriving capitalist economy, continues to hope for reunification. Established in 1969, South Korea's Ministry of Reunification has been busy planning for the day when the Kim regime collapses, and the south is flooded by millions of starved refugees who need to be fed, educated and assimilated into the strange capitalist world.

Travel in the 'evil' lands

Iran, Iraq and North Korea have experienced war and deprivation, and continue to suffer from these. But if that's all you know about a particular place, then when you imagine it, it's easy for prejudice to fill in the gaps. Many westerners likely picture Iran as a grim place with terrorists lurking behind every bus stop. Of course, the reality is that life on the streets of Tehran is much more like life in Bangkok, Shanghai or Mexico City — people are getting on with their lives, worrying about the mortgage, and hurrying home through traffic so that they can catch the latest installment of their favourite sitcom.

Without a willingness to travel into the lands we most fear, our misunderstanding deepens. Disputes between leaders mean that we as individuals start to hate and fear entire countries. And it's only when you enter — be it in person, through literature, or through conversation — and shake the hand of a local, that the shroud of misunderstanding falls away. So, let's take a peek behind the curtain.

Inside the hermit kingdom

On 2 January 2016, a young man named Otto Warmbier was detained at Pyongyang Airport when attempting to leave North Korea with the rest of his tour group. His alleged crime was the pilfering of a propaganda poster, which he apparently meant to take home as a souvenir. After a show trial which horrified the world, Warmbier was eventually released to his family in America. Eighteen months had passed, and the young man was utterly broken in body and spirit. He died shortly after returning home, at the age of 22.

This is the face that North Korea shows to the world. The incomprehensibly brutal regime that keeps its citizens living in constant fear, always on the point of starvation. The hermit kingdom, a place of incessant propaganda where, in the words of T.H. White, 'Everything which is not forbidden is compulsory.'

Just a few months before Otto Warmbier's ill-fated expedition, an earlier tour group crossed the border from China on the road to Pyongyang in late 2015. My friend Harry Cunningham was among that group. Before the visit, he, like most westerners, had seen North Korea portrayed in the media as weird beyond all reason. Documentaries — like Vice Media's famous 'travel guide' — revel in grocery store windows stocked with fake goods, and stooges apparently walking the streets of Pyongyang intending to trick tourists. He saw none of that.

'This is the sort of thing that really pissed me off when I arrived,' says Harry, 'because we saw none of it. There were no "fake" grocery stores. We saw many real ones and yes, the food was real and normal people were buying it. There were no actors. From a micro or day-to-day level, Pyongyang seemed entirely normal. Even the more interesting sights like the communist symbols and portraits of Kim Il-sung and Kim Jong-il everywhere became almost normal after a few days, not too dissimilar to seeing pictures of Mao in China, or seeing war memorial statues in various forms around most of the Western world.'

While in Pyongyang, Harry and the other tourists rode the subway, where they saw children frantically completing their homework on the way to school. Locals who they spoke to were relaxed and friendly. Yes, they had heard of McDonald's. The biggest shock, in fact, was how normal it all seemed.

'I know that this version that the media pushes, of how crazy and bizarre and unsafe North Korea is, is sexier than what I am describing. A documentary about people living in fear and paranoia, brainwashed and all as actors in a staged world sounds more interesting than a documentary about the relatively normal lives of millions of citizens of a large Asian city going to work or to school.'

And what of the gulags, the human rights abuses, the mass executions? After a few drinks, Harry plucked up the confidence to ask his North Korean guides about this. 'Their answer to that is that, the more people who come and visit and see the DPRK for what it is — a fascinating yet in a lot of ways normal place — the more the West will be able to empathize and understand the DPRK at a human level, rather than as mindless subjects of Kim Jong-un.'

Real travel goes beyond politics. A government can be evil, but citizens are mostly just living their lives. They don't want to be feared. They want to be understood.

No one goes to Iraq

Kumeait Ali was smiling. During my brief time in his company, he was normally smiling, dispatching drinks with a friendly ease at a small bar near St Sophia's Cathedral, in the city of Kiev. Being an Iraqi stand-up comedian working in Ukraine was an ... unusual occupation, but Kumeait took to it with relish. His country was seen as scary, extreme, even evil,

and it was a delight for him to show a different face
of Iraq.

About eight years earlier, Kumeait had moved out
of his home in Baghdad and come to Ukraine to study.
For the last three years, he'd lived in Kiev and mixed
with a largely international circle. There were other
Iraqis in the city, but they took their religion seriously
and kept to themselves. Kumeait, the son of a liberal,
US-educated Baghdad academic, preferred to mingle.
But Iraq is a tough country to be from.

Often, Kumeait found himself in a situation where
he was out with a new group of international friends.
Everyone would introduce themselves. Someone
would say, 'I'm from Tokyo' and people would respond,
'Oh, cool! I've always wanted to go there.' But when
Kumeait said he's from Baghdad ... well, the reaction
was quite different. People got tense. Uncomfortable.
Awkward silences often followed.

As an outgoing person and a stand-up comedian
living far from home, Kumeait tried to change people's
attitude. But he wasn't afraid to acknowledge the
reality behind some of the common perceptions.
'They hear that there is war 24/7. People getting killed.
That's kind of true. Baghdad has had many tough times.
Especially when I lived there. Now it's better.'

Still, even though Kumeait had no illusions and did
not put his own country on a pedestal, he got annoyed
by the exaggerated perceptions people held. 'We don't
lock women at home. We don't beat women. We have

a really nice civilization. Especially in Mesopotamia. Go to Baghdad or Iraq, and people will see it is more civilized than what they have in their minds.'

While adventurous travellers can go to Kurdistan, Kumeait had never met a westerner who had travelled to Baghdad. Wikitravel advises against all non-essential travel there due to 'the extreme risk of terrorism or kidnapping'. So, unable to invite friends over for a visit, he became something of an unofficial ambassador for his country.

Yes, Iraq may be a scary place, but Iraqis themselves don't need to be seen as scary people. Kumeait wouldn't propagandize or gloss over the reality. Instead, he would just be himself, and put people at ease through his personality. 'People know me. They become my friends. Now, when they see a person from Iraq, they see that maybe he is very nice too.'

At the US embassy in Tehran

The graffiti on the walls told a grim story. One banner in English and Persian read, 'We will make America suffer a grim defeat'. A painting showed the Statue of Liberty draped in an American flag, the statue's face replaced with a grotesque skeleton mask. Obama and Bush were caricatured. Graffiti disparaged Israel as the 'Unclean rabid dog' of the Middle East. In contrast to this hate, the Ayatollah Ruhollah Khomeini was shown as calm, wise and serene against the backdrop of the Iranian flag.

51

This is the embassy that was overrun in 1979 at the start of the Iranian revolution, and harrowingly depicted in the movie *Argo*. Fifty-two Americans were held hostage for 444 days. As tourists in Tehran, we marvelled at the hyperbolic graffiti showing the United States and Israel as partners in destruction, with Iran as the serene victim of their malevolent machines.

But we didn't have much time to linger. Piling into our car, we returned to the busy Tehran streets and fought our way out of the city. In a small satellite town on the edge of the desert, we loaded up on snacks for the journey. A local saw us, a group of eight foreigners, standing around the bonnet of one of our two cars, feasting on a selection of goodies we'd just purchased. He approached us and shook his head, speaking loudly in Persian.

He went to his car and opened the boot. We watched him carefully, eyes full of dates and flat bread. The man headed towards us, holding something large in his hands. It was a blanket. He spread it out on the city streets so that we could enjoy our picnic in comfort.

TRAVEL PRACTICE
Which country scares you the most?
Which country on the globe would you consider the most *evil*? Be honest about your own prejudices and the ingrained beliefs that you may not, until now, have questioned. They aren't your fault.

Take a moment to question how and why you came to hold those beliefs. Think about the media narratives you have been fed. And then ask yourself a few questions.

- Do you know the backstory: how did this country come to be this way? As a starting point, Google who ruled the country and what the circumstances of the people were 50 and 100 years ago.
- Is this context evident in the media narratives you have been exposed to, or are the headlines and news stories focusing only on the present, as if the country has always been this way?
- If you had gone through the same experiences as citizens of that country, how might your world view be different?

Honestly assess your own knowledge of this country you have feared or hated. Then, make plans for a trip. You don't have to jump on a plane — many of the countries we are taught to fear are genuinely dangerous or, in the case of North Korea, difficult and expensive to actually access. Others, like Iran, are surprisingly easy and accommodating.

But physical travel is not the only way to go below the surface. As we will explore later in this book, it is possible to travel in dozens of different lands without leaving your home country.

2.
What's in a country, anyway?

In mid-2019 social media was abuzz with the story of a 21-year-old woman who claimed to be the youngest person ever to have visited every single one of the 196 sovereign nations in the world.[1] In an interview with Forbes, the intrepid traveller said: 'I've always had a curiosity about other people's ways of life and how they find happiness.' Well, now that she has visited every country in the world and met all the people, I can only feel sorry for her — what more could there possibly be left for her to discover?

Far be it from me to criticize how someone lives their life. Visiting every country in the world at any age is certainly impressive. But to confuse this determined exercise in stamp-collecting with actual travel is to believe that adding someone as a 'friend' on Facebook is the equal of intimately knowing someone for years, listening to their stories and being there for them when times are tough. I'd rather have five friends like this than a thousand on social media I barely know. Likewise, I rate more highly the traveller who has made a thoughtful and curious exploration of four or five countries than one who prefers the bragging rights of having 'been everywhere'.

Unfortunately, this approach to travel is very much in vogue. The flaw is obvious. As we've discussed before, merely visiting a place alone is meaningless. Simply being there physically and getting the passport stamp tells you nothing, teaches you nothing, and provides no benefit to you or the locals. It contributes nothing to your understanding of the place you have 'checked into', and nothing of value to their understanding of where you are from. For travel to have a positive effect — for it to *matter* — you need to work for it. Your purpose needs to be to experience and understand, not merely to collect a stamp.

But of course, the fact that you are reading this means that I am most likely preaching to the converted. You already get this. And when you travel, you seek experience over bragging rights. However, there's one more aspect of this narcissism which is so flawed as to undermine its own purpose. Understanding this secret gives you the freedom to look at travel in an entirely new way. Realize this, and you will be able to discover hidden worlds that most casual visitors will never even notice. Here's the secret: countries do not really exist.

The 'sovereign nations' recognized by the United Nations are political inventions that often oversimplify and misrepresent that actual on-the-ground reality of where you are. This is hard to notice when you travel by air: when flying, you disappear from one commercial centre and reappear at another, where you are greeted by the swift and polished machinery of state and emerge into a buzzing metropolis, most likely the nation's capital. If you really want to get a feel for what the idea of a nation state is, and get a sense of how recent and flimsy this idea can actually be, then for your next adventure, make an effort to cross an international border over land. Not in safe, serene Schengen-zone Europe, but further afield, where different languages and cultures blend into one another across heavily policed and hastily defined boundaries.

On crossing a border

Borders, like nation states, are artificial. They would not exist if it weren't for those who believe that they do, and who enforce this belief with the aid of machine guns. And yet, take away the aloof guards, the snarling dogs, polished boots and sweaty, gold-toothed smugglers who haunt the frontiers between Europe and Asia, and you can still feel it when you enter a new land. At other times, however, the changes in landscape, language and culture occur independently of political boundaries. By the time your passport has been stamped, it's too late: you've already entered a new world.

The moment you cross from Turkey into Georgia the sun-scorched hills, like old parchment when the light softens, are replaced by lush green pastures. Gone are the mosques. Instead: orthodox churches, strip clubs, discount liquor stores and casinos. As a traveller the changes you observe — geographically, culturally, ethnically — can be devastatingly immediate. In these places, the idea of the nation holds true.

Between rustic Tajikistan and militarized, haunted Afghanistan lies a river. In places it surges furiously, silt-darkened water crashing onto the rocks and carving a deep gorge between the mountains. In other places it sleeps. For a confident swimmer, it would take only a minute to paddle across. Standing on a hilltop, I watched an Afghan settlement below. In contrast to the Tajik villagers who wore jeans and showed their hair and face, the women over the river were reduced to black shapes. Even their eyes were hidden. There were few buildings and little colour, save the grey silt of the riverbank and the brown of the rocks.

Workers, also in black, clung to the cliffs high above. With handheld tools they attempted to carve a road out of the face of the mountain. Watching them for long was disconcerting. I dropped my eyes back to the women below. From this distance, these hooded figures who strode purposefully among the houses seemed mysterious, remote, even sinister. They were less than 100 metres away, yet it felt as if I was watching a video from a fictional world. This border, too, felt real. Rivers are ancient barriers, and cultures can grow independently on either side. The political reality here also seemed to reflect the on-the-ground reality. But this is not always the case.

When heading north along this river towards Kyrgyzstan, you'll find an international border that is less precise, an example of how people seep between the 'official nations' and carry their culture and customs with them.

At a small guest house about 150 kilometres from the Kyrgyzstan border — but still snugly in Tajikistan — we slept uneasily on the floor. The air was thin at 3500 metres, and you wake up feeling as if you've been jogging in your sleep. In the morning our watches showed 7 a.m. It was 7 a.m. We'd changed our clocks in the capital, and Tajikistan has just one time zone. Why, then, did our host insist, as he spooned watery porridge into our bowls, that it was in fact 8 o'clock? We put this mystery among the others on our ever-growing pile, and moved on.

We were at the border between Tajikistan and Kyrgyzstan, on a mountaintop 4200 metres high. The white peak of Mount Lenin in the distance. Unsealed road of orange earth. The guards invented a few documents and accused us of not having them. After some muttering they accepted a few banknotes and waved us on.

As we waited for our passports to be prodded and stamped, I chatted to a vanload of Kyrgyzstani villagers returning home. Their leader was a young man, short and cheerful, who wore an enormous white hat that looked straight out of Dr Seuss. He asked us where we had stayed the night before. I gave the name of our host. 'That's my brother!' he exclaimed, delighted.

Crawling through the mountains and setting forth towards Osh, I went into my iPhone settings and changed from Tajik to Kyrgyz time. We had jumped forward an hour. It seemed our Kyrgyzstani host had insisted on living his life in Kyrgyz time, even though his guesthouse was well within political Tajikistan. In the minds of the locals, we had entered Kyrgyzstan several hours before we crossed the 'official' border. It would have been completely possible to experience the real Kyrgyzstan, talk with Kyrgyz locals about their customs and enjoy their hospitality and lifestyle, without ever officially entering the country.

'Where am I?' is a complicated question

Structuring or evaluating travel based on countries, and ticking off these countries on a list, orientates you to a world view that can diminish the real experience of travel. Some of the most interesting places exist not merely in one defined political space, but across many different worlds, where different religions, cultures and identities jostle for survival and supremacy.

If your focus were on acquiring passport stamps rather than acquiring experience, this entire reality would be completely missed.

That's not to say we should ditch the idea of countries altogether. Despite the many flaws of this idea, not least of all the ludicrous modern invention of nationalism, they are the primary way in which our world is organized. Because people believe in the idea of a nation, nations become real because we as humans make them so. (For a fascinating exploration of this, check out the rightly popular *Sapiens* by Yuval Noah Harari.)

The identities which we assume are with us from birth are less absolute than we realize. The country listed on your passport may only be a couple of hundred years old, if that. The ideas and values associated with that name likely did not arise in a state of perfection thousands of years ago, but were most likely invented by politicians with an agenda. It's a scary thought — the idea of where we come from is absolutely integral to how we see ourselves. Even writing these words, I feel the power that the idea of being a New Zealander holds for me, and we are proudly a young nation of immigrants. Those of us who are from ancient countries, or countries that have had to literally fight for their right to exist within living memory, hold this idea even more strongly. And I am not asking you to throw it completely out the window, but merely to take a step back and look at the idea of 'nationality' from afar, examining this idea as if it were a foreign custom.

When we start to question the legitimacy of the idea of a 'nation' it opens us up to new, exciting, and maybe even challenging ways of looking at the world and our place in it.

In Chapter 1, we explored the idea of putting a country

into context by understanding its centre of gravity. For this, we looked at how getting some rough sense of the complex dynamics of a country can help you, as a traveller, know where to look and which questions to ask. That chapter was about attempting to reconcile the many contradictions of a place in order to start building a useful idea of a coherent 'whole'. Now we're going to zoom in on a micro-level and embrace some of the many contradictions you attempted to smooth over in the exercise on p. 52.

Discovering worlds within worlds

On any given street, a number of different worlds exist side by side. Oftentimes, these worlds exist within the minds of the people you encounter. The Kyrgyzstani man who lived in Tajikistan but set his watch to Kyrgyzstani time brought his own world with him. He was himself an enclave. Other times, they can be observed in the physical world around you.

Romania is a country with a strongly Latin-influenced language that, naturally, uses the Latin alphabet. Yet when the British writer Elizabeth Harris visited a small village in the countryside, she noticed the signs gradually began to change. As she recalled in a 2019 article for Intrepid Times:

The road to the village took us through open farm-land and a number of small hamlets comprised of well-kept wooden cottages with carved porches and painted shutters. Waiting horses dozed by the side of the road, attached to carts laden with hay or timber. The place names were written in both Romanian and Cyrillic, hinting at the proximity to Ukraine.

Here, Elizabeth had found a place where the neighbouring culture was so strong it had almost literally spilled across the border, passing the European Union's border protection and rewriting the very road signs.

Different worlds spilling into each other is common across Europe. These contradictions are felt particularly strongly in the towns and communities that line international borders, but you can in fact discover these micro-worlds pretty much everywhere you visit. They exist within the minds of the people, and are often the key to understanding the fabric of a place.

To discover the worlds, it can be helpful to once again begin with your own lived experience, examining yourself and your own world as you would a stranger in a faraway country. You can then take this mould and apply it to somewhere else, filling in the blanks with your observations and interactions when you're actually there. So, think about your home town and the number of different worlds that exist on your street, on your block, in your suburb. If you're from a cosmopolitan metropolis like New York or London, then this will be easy. You probably know people with dozens of different religions, ethnicities and backgrounds. They each exist in at least two worlds: in the world of memory, tradition and culture they bring with them; in the immediate world you live in as a fellow local; as well as in the grey area between the two, the lived experience of an immigrant. Elsewhere, in smaller towns, less diverse cities and more isolated areas, these signs of different worlds may be less obvious, but they will still be there.

The history of our species is the history of inter-mingling. We are all mongrels. There is no absolute racial purity. All cultures and religions are mixtures, adaptations and interpretations of various influences that have come before. The clues in your home town may be subtle, but they will be there. What kinds of churches or temples are present? What are the surnames of the people you know? What are the schools, and who goes to which one? How long have the families you know lived there, and where were they before? If a bright traveller asked you over a drink to describe yourself and your town, what would you say? What if this same traveller asked your neighbour, your doctor, your parents — would they get the same response? What points would change? How would the focus and the emphasis shift?

You have this idea in your head of what a place will be like. Sometimes, it's exactly as you expected and other times it's more different than you could imagine. Like the time I visited the quaint medieval walled town of Lucca in Tuscany for its historic value — during the only weekend in the year when it hosts the European version of Comicon, and the population had increased fivefold! The streets were heaving, literally filled with people in costume — Batman, Spider-Man, assassins with guns, brides, Star Wars characters and anything else weird or wonderful you could conjure up. I should have been disappointed, but how could I be, really?

Judy C, New Zealand

TRAVEL PRACTICE
Count the worlds around you now

Begin with yourself. You may have been born on the same block of land on which you're now sitting with this book in your hand. But what of your parents, your grandparents, their parents? What about the Chinese takeaway around the corner? You've probably never asked yourself which province the owner came from. Ask a Chinese friend to tell you if the sign is written in simplified Chinese characters or traditional characters. If traditional, it means the family is most likely from Guangzhou or Hong Kong, a part of China traditionally more open to foreigners and an easier path of escape during the country's upheaval in the 20th century. If the sign is in simplified Chinese, they are most likely more recent immigrants from modern China. There's a story there.

No Chinese takeaway? Look to the churches. Maybe you live in a small town with only two churches and a general store. So, why two churches? Which denominations do they serve, and where did these adherents originally come from? Catholics from Italy, protestants from Britain? What happened in those countries years ago that caused such a wave of migration?

Looking twice at things you had previously only glanced at can uncover a richness of history and make your own sliver of land feel connected to events happening thousands of miles away. It's not only fascinating, but it's also great practice for getting your eye in so that when you're on your next trip, you'll be able to discover and learn more from the instant you arrive. When you tune in to these stories that constantly surround us, you'll be able to experience more from one city than your average passport stamp collector would get from 50 countries.

As you start developing this awareness of multiple worlds that exist side by side in any given space, you may find that you gradually develop the habit of travelling slower — of valuing how long you spend in one place over how many places you have been to. And when you let go of the natural but ultimately stifling need to cross the names of countries off a list, to constantly update your mental scorecard, then you are closer to getting out of your own way and becoming more in tune with the immediate experience of travel.

3.
On identity

If you think you are an American, British, a Kiwi, an Australian, French or German, or to any degree allow your nationality to influence your identity, then you may owe something of who you are to a group of pampered diplomats who met in Central Europe almost 400 years ago.

The year, as they say, was 1644, and representatives from 194 states had gathered to talk about ending the Thirty Years' War — the series of bloodthirsty conflicts that had sprung up across Europe as a direct or indirect result of the Austro-Hungarian empire attempting to force Catholicism on Bohemia. Negotiations took the better part of four years. This is hardly surprising: the first six months were dedicated to deciding who would sit where, and in which order people would leave and enter the various negotiation rooms.[1] But the resulting success, what we now call the Peace of Westphalia, had far-reaching effects. It's one of the reasons why you have a passport, and why it says Germany or Australia (or wherever) on it.

Westphalia put in writing the idea that each territory would have sovereignty over what happened within its borders, and that other rulers would have to respect this. Every nation state, big or small, controlled what happened in its own land. Today, this principle is known as Westphalian Sovereignty and it is enshrined in the United Nations Charter. While Europe was still very much a land of empires until after World War I, Westphalia seeded the idea of nationality. Before Westphalia, people didn't really think in terms of 'countries' or 'nations'. Your identity was tied to your family, your tribe, the feudal lord who came around once a month to claim 50 per cent of your chickens. The person in charge was most likely seen to be chosen or legitimized by God, and that was pretty much it. After Westphalia, identity began to be tied increasingly to geography.

The invention of nations required the invention of national identity. Some geographical slices already contained people who shared a religion, language and ethnicity, but for the most part national languages as we know them today did not really exist. People in 'France' did not all speak 'French' — they spoke a regional dialect that may have been more or less incomprehensible to folks in the town 15 kilometres over. That 15 kilometres would, after all, have been a long and laborious journey. For the most part, your village would keep to itself, and the idea of 'being French' wouldn't really have meant that much to you. But then, gradually, it did. The rise in national identity was in part the result of technology; the printing press created mass media, which necessitated a common language and allowed for the telling of national stories, myths and propaganda that bound previously disjointed people together. It is, in part, also the deliberate result of government policy: even now in much of the world, the history taught in schools has a distinct bias towards the forming of a coherent, if not absolutely accurate, story about who you are as a citizen of where you are from.

This may sound like it belongs to ancient history, but it does not. Nations are so young that you can probably see evidence of the creation of national myths even in your own lifetime. Think back to your schooling. Even if you are from a liberal country where you are not required to look at smiling portraits of your Dear Leader at every bus stop, you have almost definitely been the unwitting recipient of some form of mildly nationalist propaganda ever since you first entered formal education.

> The most obvious impact of the nation state is the creation of a uniform national culture through state policy. Its most demonstrative examples are national systems of compulsory primary education that usually popularize a common language and historical narratives.[2]

Think about the events you were taught in history, and how your country was likely cast as the hero. Think also about how, when you reflect on wars and conflicts of old, you might be inclined to use the pronoun 'we' when describing the exploits of your own country.

'We fought the French at the Battle of Waterloo.' 'We fought the British in the War of Independence.' With newer countries like the United States and New Zealand, these national 'founding stories' are clear and easy to trace. With older nations like Britain and most European countries, their stories may be murkier, and it's all too easy to look back at history with the labels and attitudes we have today.

The idea of defining yourself based on what your passport says is very new indeed. And what this national label says about you is unlikely to be accidental. What it means to be 'Irish', 'American' or 'Australian' has been carefully debated and crafted by bureaucrats and politicians, and the artists and writers who seek their approval and business. In a modern democracy, we all get some hand in shaping and updating this identity. But even in fiercely individualistic cultures like the United States people still subsume a great deal of their own personal identity within the identity of the country they come from. This, at its worst, can grow into the political ideology of nationalism, where warped myths about what it means to be a 'real' English, Polish or French person are used to marginalize and exclude foreigners who do not fit this definition, never mind that this supposedly pure national identity is a modern invention.

Leaving home, in body and mind

The fact that you likely derive some sense of identity from the country where you were born, the lands your parents come from, and the title on your passport does not mean you are a raving nationalist. Pretty much all of us get a sense of identity from our country. In the modern world, land of internet and jet travel, this unit of definition makes sense, as we explored in the previous chapter. Thanks to mass media, you probably do share a lot of commonalities with others from your country: not just language, but a sense of the events that are important to you, as well as of course cuisine, music and other aspects of culture. Unless you're unusually politically active, you probably don't think about your national identity on a daily basis. But travel for any long period of time and it will start to come more and more frequently to mind.

It's always the same: whenever you meet a new friend at a youth hostel or travellers' bar, the first question is always 'Where are you from?' And in answer to this, you are not expected to give a street address. And unless you're from New York or London, you're unlikely to even name a city. What most of us say in response to this question is the country that is listed on our passport. And on this alone, we are judged. It becomes like a label on our forehead. And we start to believe in it. Start to lean into the stereotypes, start to feel pride at its successes and shame at its failures. We start to become, in this foreign land, the representative of the political and geographic entity in which we entered the world.

This process can be a surprisingly powerful one. It was only after I left home and spent a lot of time abroad that I started to really identify as a New Zealander; before then, the idea of being a 'Kiwi' hadn't meant that much to me, consciously at least. Being overseas can have a contradictory effect when it comes to understanding who you are and your place in the world. On the one hand, the exposure that travel gives you to different perspectives and different narratives on the world can allow you to look at history in a new way and question some of the national myths and assumptions you grew up implicitly believing. And on the other hand, travel often forces you into a position where, surrounded by strangers, you are cast as the sole representative of your homeland. This, combined with homesickness, can cause you to emotionally lean into these very nation-building tropes and stories that you are simultaneously beginning to question. Travelling overseas made me feel more 'Kiwi', even though I was seeing the world through an increasingly global and varied lens.

The invention of identity

As we explored in the previous chapter, nation states are imperfect. Their borders are rough. Within them and around the edges are groups who claim their own identities. These could be connected to religion. To ethnicity. But they are, to some extent, tribal. They are unique, but they also separate and divide. There cannot be an 'us' without a 'them'. Particularly in the developed world, different identities have started to come to the foreground. Identities related to gender and sexuality are in vogue at the time of writing. Nowhere is this more fraught than in the United States, where so-called 'identity politics' has become the defining feature of political debate. People define themselves by a certain characteristic, normally race, gender or sexual orientation, and make political decisions based upon what they perceive to be important to the specific interests of that particular group.

Unsurprisingly, none of these groups in the United States (or anywhere) today are particularly happy, as Amy Chui writes in *The Guardian*: 'Whites and blacks, Latinos and Asians, men and women, Christians, Jews, and Muslims, straight people and gay people, liberals and conservatives — all feel their groups are being attacked, bullied, persecuted, discriminated against.'[3] Now, while nations are political and narrative constructs, gender, race and sexual orientations are real. Right. You're either white, black, brown, gay, straight, bi or … well, whatever you are. That's that. You didn't choose it and society didn't invent it. But of course, society did invent it. Sexual norms are very much the product of religious orthodoxy — the ancient Greeks didn't care who was screwing who, man or woman. Likewise, while people of different races clearly look different, this superficial distinction is basically meaningless on a genetic level.

73

It was just a convenient way for people to divide others.

Unfortunately, the racists and bigots have gone a long way towards actualizing their prejudices. A black person in the United States will, in a great many ways, experience life differently to a white person. Their 'lived experience' and the actions of bigots make a fabricated and nonsensical difference real. In order to address this, these minority groups bind together out of necessity.

Travelling overseas won't change your nationality, your favourite sports team, the colour of your skin or your sexual preferences. It may, however, give you space to wear these identities more lightly. You'll still be the same race, gender and sexual orientation you are now no matter where you go and who you meet, but the meaning this has to you might change based on the experiences you have and the people who choose to share their own experiences and stories with you.

The power to understand and be understood

One of the many toxic aspects of the 'identity politics' debate is the common pattern you will often see in op-eds and Twitter posts: 'You could never understand my experience as Y because you are an X', 'You could never understand my experience as a gay black woman because you are a straight white man'. You probably don't have to think too hard to see some variation of this formula. There is certainly some truth here. No, I do not know what it's like to be gay, black or a woman —

certainly not all three. There are life experiences, struggles, challenges and pains that are unique to certain groups. These experiences should be respected. But this should not be a tool to shut down conversation. Far from it. This should be the spark for conversation.

I don't know anything about what it's like to be a 70-year-old Sichuanese farmer. But when I stayed for a couple of nights in rural Sichuan with a 70-something-year-old farmer and his family, he (in a heavy Sichuanese dialect that was almost impossible to understand) did not get offended when I asked him about his life experience. He didn't say 'You're a 20-something New Zealand student, you wouldn't understand.' The whole point of language, of storytelling, of communication, is to take one person's experience and translate it so another person can understand something of it. Travel exposes you to people with vastly different life experiences, vastly different centres of gravity, who have had radically different experiences of life. And your job as the traveller is to listen, to ask questions, and to be informed. When we travel, we look with curiosity upon those who have different life experiences than us, and treat others who view us with this curiosity and interest with indulgence and patience. But when home, it's too easy to fly the flag of moral outrage and get that holier-than-thou kick out of saying 'you couldn't possibly understand'.

Trying to understand and appreciate what life is like for someone with radically different experiences is called courtesy when you travel. But back home, people can label it 'cultural appropriation'. Sometimes, it feels as if there were a conspiracy designed to keep people divided. And if you thought this, you'd of course be right. Politicians and political consultants strategically use issues to drive wedges between different groups. When the opposition is squabbling, it's easy to stay in power. Our fears and frustrations arise both from our legitimate experience but also from the invented identities imposed on us by greater groups. Travelling allows us to take a step back from the noise, and understand from a distance forces, often shadowy and political, that play an outsize role in shaping our own identity. It also allows us to place this identity in context. To filter our own narrow experiences through the vastness of experience and perspective that you will encounter on the road.

TRAVEL PRACTICE
Lose yourself

One of the many joys of travel is the ability to cast yourself as the hero of your own story and experiment with new sides of your personality. While the folks back home might have seen you as shy and reserved, away from these expectations and associations you are free to break out of old patterns and try something new. If you embarrass yourself, so what? You're off to the next city tomorrow, try again. Travel can allow you to craft a more accurate sense of identity by exposing you to more data about the world and providing the opportunity for you to shed the baggage that has been imposed on you. It can help you reconnect with parts of yourself that may have been buried or stifled due to our recent confinement. And it can also help you do one better: not simply to *find* yourself, but to *lose* yourself. A scary concept, but a profoundly liberating and empowering one if you take it just far enough.

Try this thought experiment. You are alone in a foreign country where they speak a different language and have a vastly different experience of life. You arrive late at night and are picked up and taken straight to your hotel. You sleep off the jetlag, freshen up, and in the morning you step out onto the street. Immediately you are entranced and overwhelmed by the sights, sounds and smells that surround you. Everything is different! The stress, the language, the manner of life that you see everywhere you turn your head. But what do the locals see when they look at you? Would they think you are ... tall? Rich? Lost? Well dressed or poorly dressed? Would they assume you are a Christian, or Muslim, or would religion not cross their mind? Make a list of five things. And then compare this with the truth of who you are.

4.
It's not just about you

You can't truly connect with your experiences as a traveller if you are too focused on yourself. But there's a simple trick to taking yourself a little less seriously.

The students around me openly pointed and stared. Many snapped photos with their smartphones. Some tried to be subtle about it, pretending to take selfies or quickly stealing a picture as they walked by. Others thrust their phones right in my face. A few were polite and asked for permission first. Many shouted openly, calling to each other 'Look at that foreigner! He's so tall. Must be a Russian. No, he looks like an American. Wow, they're so white!' They spoke in Mandarin, but by this stage I had been studying long enough to understand pretty much every word. I kept on walking.

As a tall, blond student of Chinese language at a second-tier university in a second-tier Chinese city, I quickly got used to being gawked at like an exhibit in the zoo. At first it's amusing. Then it's annoying. Then you don't even notice. After a while, it becomes quite liberating. You accustom yourself to the giggles, to the

blank stares when you attempt to communicate, to
the amazement when you get the language right, to
the occasional rudeness and more frequent
confusion.

Even buying a bottle of water or taking a short trip
becomes an elaborate exercise involving an online
dictionary and half a dozen attempts to pronounce
the words you've learned in class in the way that a
Sichuanese shopkeeper or taxi driver would understand.
When it works, you feel like a superhero. Here you are
in a city you hadn't even heard of a year ago, speaking to
locals in Chinese! When it doesn't work, you feel like an
idiot. Until ... you don't. Until the feeling is so familiar
that it doesn't affect you any more. The embarrassment
is gone. It doesn't matter who is watching.

There's the terror that first precedes every interaction.
You anxiously check the vocab. Practise the words in
your head. Walk into the store or pick up the phone, and
pray that the interaction doesn't go off-script. Inevitably,
it does. Sometimes because of the script, or sometimes
your brain just freezes and reaches for the first word or
sentence it can find, no matter the language, context
or logic!

Those who succeed soon learn not to take themselves
too seriously. To be comfortable with the awkwardness,
to embrace the embarrassment as a healthy part of
the learning experience. After my six or so months
in Chengdu, Sichuan, I also learned patience. Never

again would I get frustrated — even internally, even in my own head — with a foreigner struggling with the English language. Take your time. Take 30 minutes to understand my order for a black coffee at Dunkin' Donuts. I've been there, mate. I have been you, I know how it feels. Take all the time in the world.

The goal of travel is not to find yourself; it is to lose yourself

Everyone should have the experience of being an alien, of struggling to keep your head above the water in a sea of foreignness, unable to even read a fast food menu, a street sign or a bus timetable. To be utterly dependent on interpreters for your survival, but to venture out on your own and, bravely, word by word, decipher the world around you as if from scratch. As vulnerable as a baby, but also as curious. An empty sponge and, at the risk of breaking the world record for most metaphors in one paragraph, a white canvas, ready to be splashed with the colours of the world.

When we travel we bring with us a suitcase full of clothes, and carry on our backs the attitudes, beliefs, ideas and stories we use to define ourselves and filter the world around us. These mental constructions likely serve us well at home, where we exist in a community defined by clear rules and conform to the deeply ingrained expectations of those who rely on us, and we on them. You probably define yourself by a set of statements, many of which are likely in the negative. 'I am sensible; I don't make a fool of myself in public.' 'I don't take stupid risks.' 'I am confident and in control.' 'I don't eat meat.'

When you travel, these artifices of home life hold you back. They become barriers to the immediate experience of where you are. Before going to China, I did not consider myself massively outgoing; coming of age in the boisterous Otago University student culture in Dunedin, New Zealand, I had often clung to the fringes, held back from fully immersing myself in the student experience due to an excess of self-consciousness. But in China, I eventually learned to let this side of myself go. It's easier to overcome such inhibitions when you are abroad, particularly if you are travelling for a long time. When you aren't surrounded by people who already know you and see you through the lens of who you have been around them, you have more freedom to reinvent yourself. By doing so, you not only have the opportunity to improve yourself as a person, but also to get more out of the experience of travel as it is.

To leave your old identity at home is a terrifying thing. You probably don't even remember what life was like before you had it. Rather than seeing the more constricting aspects of yourself as obstacles to be overcome, many of us see them as permanent and indelible aspects of ourselves. These characteristics have served a useful purpose in the life you live. You did not develop your self-consciousness, fears and beliefs for nothing; they keep your behaviour acceptable within the tribe you are in.

Leaving all of this behind while still at home risks making you feel like a freak, an outsider who transgresses the rules that everyone else plays by. Someone who is different, and so risks being perceived as *wrong*. But when you fully open yourself up to the travel experience, you'll understand that once you step out of the airport, an outsider is exactly what you are. In much of the world, you probably are something of a freak, an anomaly, one who is *different*. And when you realize that's okay, you open the door to a far deeper and more profound experience of wherever you are.

Self-consciousness is the opposite of travel

When was the last time you travelled out of your so-called 'comfort zone' — the last time you were in an unfamiliar environment, a foreign culture, a strange world? Now ask yourself ... were you really there? Was your mind focused externally on the sights, sounds, smells that surrounded you? When you spoke to locals, did you really listen to their stories?

Self-awareness is essential when you travel, for your own safety as well as for the value of the insights it provides. Imagining how you are perceived as an alien in a foreign country is a fascinating thought experiment that can provide deep insight into both where you are and who you are. But self-consciousness is different. It is the experience of being unable to fully experience. It is the inability to focus on others because your *self* keeps blocking the view. It is a condition we all experience to some degree or other. And in the selfie-obsessed world that we live in, this self-affirming narcissism can drive us to value a travel experience not because of what it tells us about the world, but because of how 'cool' we think it makes us look to others.

This is not to rant about the social media age. Humans have always been driven to tell stories, and adventurers since at least the time of Herodotus have sought to use their journeys to seek the adulation of the crowd. Particularly for those of us born with the disease of literary aspiration, there is always a part of the mind that is separate from the immediate experience, seeking to structure it into a narrative that can later be retold in verse, conversation or Tweet form.

But all of this is to some degree antithetical to the travel experience. Surely the point of being somewhere is to *be there*, not to be concerned with the external gains that being there will bring. Not to be there in order to use the world as a mirror. But merely to be. It is well known that the process of observing a phenomenon changes the very nature of what is observed. In other words, it is impossible to fully do something while also being fully aware of what you are doing.

I went into a small henna tattoo parlour off a busy street in Goa, India. I was the only one there, and I

> *chatted with the woman there as she drew a beau-*
> *tiful design on my hand. She told me her story,*
> *how she was one of a family of eight siblings and*
> *travelled there every year during the summer peak*
> *season. She said it was rare to find a tourist who*
> *actually showed a real interest in her story, but*
> *talking to her was one of the highlights of my trip.*
>
> Joanna N, Poland

It's easy to take yourself too seriously and believe that the derision of a stranger actually matters, or to value your sense of the world more highly than the immediate experience of the world as it is. But this attitude stifles your enjoyment of travel. It takes your focus off your surroundings, off the adventure, and puts it back on yourself. Self-consciousness is incompatible with true travel. You cannot lose yourself in a strange land if you are constantly worrying about what others may think. You cannot truly be there if you are too busy being you.

The experiences of the Australian writer Fiona Davies perfectly capture this. Fiona travelled to southern India to observe a days-long religious festival in the city of Madurai. Standing on the side of the road, she observed procession after procession of luridly dressed locals dancing and clapping to the rhythms of ancient religious music. During her second or third day, she was suddenly grabbed on the arm by a young man in one of the percussion groups passing through, and found herself suddenly in the middle of the procession. This is how she described her experience as part of an article for Intrepid Times:

> I am a terribly self-conscious dancer and usually loathe 'the circle' and the resulting surrender of free will it elicits. This time though, the circle is made up of half a dozen enormously excited Tamil boys, their eyes full of encouragement and anticipation as they propel my body into the centre of the action. There is nothing I can do. I begin swaying my hips and twirling my wrists, even going for a bit of side-to-side neck action and a few shoulder pops for good measure. My Bollywood dance move repertoire exhausted in minutes, I exchange smiles of gratitude with the boys and retreat into the throngs of bystanders, half in disbelief, giddy and elated.

Notice how self-consciousness evaporated as Fiona opened up to the wonderful opportunity of the moment. This one paragraph describes the experience of real travel in action — a traveller willing to shed their inhibitions, shed their restrictions, shed what may have been true about them for a lifestyle, in order to fully participate in the moment they are in. It's about accepting that you are part of something bigger than yourself. It's an easy thing to talk about, but it's a harder thing to actually accept. After all, we as humans are naturally predisposed to focus on the 'I'.

The first person pronoun

As the editor of Intrepid Times, I review several submissions every day from experienced and aspiring travel writers who want to share their stories with the world. The urge to tell a story is a deeply human one. The storytelling impulse shapes culture, and the urge to share information is what makes us human. But not all stories are of equal value. As an editor with an ever-growing pile of submissions to sort through, and limited time and resources in which to do so, there's an easy test I apply to the stories that come into my inbox: the word 'I'.

When writing in the first person, it's perfectly natural and okay to throw in the word 'I' every now and then. You will most certainly see it crop up in this book. It's a legitimate narrative technique and the use of it is not an error. But excessive use of this word without appropriate context reveals where the mind of the writer truly is. A writer who submits an 800-word travel story in which the word 'I' appears as many as 50 times typically isn't really writing about the destination that they are in. They are writing about themselves. They may well be a lovely and interesting person, but unless they are a president or a celebrity, chances are the reader doesn't care about *them*. They may care about the experience this writer has had in a foreign land, but if the writer is too preoccupied with themselves, then this story will never truly come through.

The best travel stories we receive do not always come

from experienced writers. They aren't always from people who have studied under great authors and who have a binder full of impressive clippings. They come from people who *travel to connect*, rather than those who merely travel to be seen. They come from people who open themselves up to the immediate experience, whether it's meeting an old man in Sri Lanka who tells them stories of the civil war, travelling in Iran as a woman, following in the footsteps of Fredrich Nietzsche in Switzerland, or teaching English in Vietnam.

Often, I've had to reject stories that would have been fascinating in other hands simply because the writers could not get over themselves. It is possible to journey through some of the harshest, most remote and most fascinating regions of the planet without ever truly being there, and while being focused on yourself the entire time. The resulting stories tell of fascinating places, but they do not make fascinating reading. The reader is alienated because the writer doesn't care about them. They don't really care about where they have been, either. All they care about is themselves. This is a poor approach to both travel and writing.

Far better are the stories that we receive, often from first-time writers, about a specific travel experience the writer found moving and powerful. The best travel stories are often written simply to capture and communicate the core of this experience. The focus of these writers has never been on themselves. It has been on the places they have been through and the people they have met while there. They are conscious and awake to the immediate experience of travel. This alone is more valuable than any writing course on the planet.

TRAVEL PRACTICE
Getting over yourself

In order to fully connect with the experience of travel, to fully connect with other people, perspectives and cultures, you must lose yourself. And in order to lose yourself, you must first lose your fear of the crowd. But how?

Well, there's one shortcut that may help you shatter the chains of self-consciousness and feel the excitement as well as the fear of living life outside the rules and norms that you conform to at home. This exercise is something you can do anywhere, even in your home town. It will give you all the feelings of awkwardness, fear and shame that failing at speaking a foreign language overseas gives you, without the cost of the plane ticket. And then once it's over, you will experience the same feelings of liberation and relief. You will have transgressed in public, incurred the derision of your peers, but the sky will not have fallen down upon your head. You will have learned how to take yourself less seriously.

Here's how it works: go find a chain store, like a fast food restaurant, a place where the staff pay little attention to each customer, and people are churned through. Line up, and when it is your turn, with a completely straight face and acting as naturally as possible, ask for something that you know for a fact they do not have. Ask for a glass of chardonnay at Burger King. Order a Big Mac at Domino's. Don't smile. Don't make it seem like a joke. Act as if you were asking for the most obvious thing in the world, and feeling the ice trickling down your spine as you do so. I have done this. It is *terrifying* — hundreds of thousands of years of human evolution pushing you to conform have made it so. But do it anyway. And when your request is rejected, thank the cashier politely and walk away.

It's impossible to meaningfully connect with where you are or the people you meet there if you're constantly in your own way. This exercise — silly as it sounds (and is) — will help you get over yourself and be more present wherever you are. Try it. And then when you're next overseas, about to speak a foreign language, about to try to buy train tickets in French or order a taxi in Italian, the experience will be that much less frightening.

5.
When travel isn't a choice

He was sitting at the table chatting to my friend when I came back from the bathroom. Dressed like a rock star, perfectly coiffed hair, contoured beard, a stud in his left ear, golden buckles shining on his blazer. He introduced himself as Enzo, from Iran. 'It's an Italian name, I know. It's not my real name. But everyone here knows me as Enzo.'

Enzo chose this incongruous club on the wild streets of Berlin as his stage because there was nowhere else on this planet where he was permitted to sing. Couples straight, gay and everything in between leaned against graffitied walls, chain-smoking cigarettes and knocking back beers from the bottle. We sat around a sticky wooden table and let Enzo share his story.

In Iran, Enzo had taught at university and sung in underground clubs, illegal but popular, until the secret police acquired a video of him kissing his girlfriend in public. This was enough to force him to flee the country and apply for refugee status in Germany. Now in legal limbo, he was unable to work, unable to access his resources from home, unable to take to the stage in any legitimate house of music and put his talents to use.

My friend and I drank beer as Enzo talked. He would have a drink after his song, he said. Alcohol dried out his vocal cords. 'And the cigarettes?' He shrugged. He wore a cross around his neck in honour of his mother, a Christian. His father was a Muslim who had died some years before.

Living on a Prayer wobbled and raved off the stage, crashing into the walls and drowning out our conversation. Enzo took a drag on his cigarette and continued. 'All my life I wanted to sing opera. But it is illegal in Iran. To have opera you need a man and a woman to sing together. And this the government will not allow.'

Reaching for a packet of tissues in his pocket, Enzo dried his hands nervously. His first song was approaching. He made his way to the stage, jostling among the revellers. The MC was a black man dressed in a green leotard. A master of the camp stage voice, he introduced each new act with giddy excitement, replete with rolling r's and fluttering vowels.

Two women in their late thirties left their beers and shuffled stiffly onto the stage to perform a respectable version of *99 Luftballons*. A clearly wasted punk rock girl in a ragged hoodie belted out Wheezer's *Say it Ain't So*, and a Vietnamese man yodelled his way through a country ballad.

It was close to midnight. The crowd was drunk, accustomed to singing along loudly to the familiar party hits. Then Enzo took to the stage to sing ... opera? As the opening bars of *O Sole Mio* began, the crowd became restless. People muttered, some giggled.

Enzo stood on the beer-soaked stage as if he was performing at the Royal Albert Hall. He hit the notes strong, keeping a straight face, singing with passion. As he lilted and rolled through the song we forgot about the camp club and its slippery floors, the gleeful MC and the jittery performers who had come before. We all just stood in silence, taking it in.

When the song finished, the few who were left sitting down leapt to their feet in rapturous applause. Enzo flashed a shy smile at the audience and handed back the mic. As he walked past I clapped him on the back. 'Now you can have that beer, mate!' He grinned but shook his head; he would be singing *Nessun Dorma* later that night.

A stranger is an opportunity to travel

In some ways, Enzo is one of the lucky ones. He has a talent that draws people in. The power of his voice and presence won over the previously sceptical crowd. Everyone in that bar would become his friend. No one there would later be likely to go home to Tweet anti-immigrant hate messages, scrawl abusive graffiti on city walls, or lend their support to one of the many thriving European far-right parties.

While those of us from the privileged world travel

because we want to, there are those who travel because they have to. Some are escaping war and persecution. Some travel because the economic situation at home makes it impossible for them to feed their families. Others are looking for somewhere they can live safely, date or marry according to their nature, and express themselves without fear of arrest or persecution. It is increasingly common to hear people questioning some or all of these reasons. 'Economic migrants' has become a dismissive term.

There are many legitimate problems with mass migration in Europe. Not least of all demographic, where males between eighteen and 34 make up a disproportionate number of European asylum seekers — and many of those from Middle Eastern and North African countries hold attitudes towards women which are drastically different from those in Europe.[1] Countries like Norway are making heroic efforts to educate these migrants into the sexual and political norms of Europe, with classes for refugees who are unaccustomed to seeing 'women in skimpy clothes drinking alcohol' on how to behave appropriately.[2]

It is of course a given that people who move to a new country should respect the laws and cultural norms of that country. The political equality and social freedoms that make Europe so attractive in the first place were hard-won by centuries of thought and struggle, and governments are right to insist that they are upheld. But just as, when we travel, we expect to be greeted with some level of open mindedness, perhaps even warmth, the same benefit of the doubt could better be applied to refugees. Instead of reacting with anger, hostility and absolutist far-right politics, those who feel threatened by mass migration to Europe could instead try to reach for curiosity. We could view strangers as an opportunity to travel.

'Travel is fatal to prejudice' — but only when done right

Mark Twain famously wrote:

> Travel is fatal to prejudice, bigotry, and narrow-mindedness, and many of our people need it sorely on these accounts. Broad, wholesome, charitable views of men and things cannot be acquired by vegetating in one little corner of the earth all one's lifetime.

The moustached raconteur was right about a great many things, and seems to have been right about this as well, at least according to a 2013 study in *Social Psychological and Personality Science*.[3] In a survey of 700 people across five studies, the researchers showed that, broadly speaking, the more places someone had visited, the more trusting they were likely to be of others. They concluded that the breadth of travel experience was key. Giving someone more exposure to people who look, live and seem different allows them to conclude that different is not always scary.

Much of this book argues for the case of depth over breadth, but this study shows that we need both. For a white person from England, travelling only within the majority white and majority Christian nations of Europe will only provide so much opportunity for your world view, and your view of others, to be challenged. Travelling in a country where most people look different to you, worship different gods and follow different customs is, on the other hand, a reasonably sure-fire way to lose your fear and suspicion of 'the other'.

Having travelled in Iran, I immediately reacted with curiosity upon meeting an Iranian in Europe. I have never been to Syria and am unlikely to visit anytime soon, but I have met Syrians abroad and would welcome an opportunity to learn more about that country. When I was in Morocco, I was treated for the most part warmly by the locals I got to know. It is very unlikely that most white European nationalists have taken the time to travel in Iran, the Middle East, North Africa or indeed any Muslim nations (which also include Malaysia and Indonesia, for a start). If they had, and experienced the tradition of hospitality, made new friends and discovered that they were treated kindly as foreign guests with their own strange and unfamiliar customs, they would find it harder to fear and demonize these same people back home.

It's hard to hate something you don't fear. Much

fear is based upon the fact that something is unknown — as my childhood hero, the bearded wizard Albus Dumbledore, put it, 'It is the unknown we fear when we look upon death and darkness — nothing more.' When you travel to a place, it stops becoming like the dark and unknown night, and instead becomes a series of rich and textured memories. When I think of Iran, I think of chatting with locals by the Caspian Sea, the taxi driver who gifted us his prayer beads, the taste of fresh lemon juice seemingly added to everything. But for people who have not had the experience, it is easy to fill in the gaps with visions of what we fear — our human natures and our media diets make it so.

And yet somehow, this magic transformative effect of exposure to foreign cultures seems not to work for many of us when we're at home. We do not allow it to. When we're home, we turn off the curiosity switch. We tolerate our foreign neighbours, rather than being curious about them.

Forget tolerance; let's celebrate curiosity instead

The very word *tolerance* implies that foreign cultures and people are a source of suffering to be endured. You tolerate a noisy neighbour, a dawdling co-worker, a car that has seen better days. You don't 'tolerate' a romantic night with your partner, a beautifully crafted martini or a new episode of *Game of Thrones* — you enjoy it (except for the last episode, of course; don't get me started).

What we need towards migrants, refugees, immigrants, in other words the *strangers* in our countries, isn't tolerance, it's interest. It's curiosity.

We explored earlier in this book how no one particular place is as simple as it seems. Everywhere you go, there are many worlds living side by side. But how much time have you really taken to explore these varied worlds in your home town? As is now obvious, it's not enough simply to live in a place where there are a lot of foreigners in order to learn about other cultures and overcome our prejudices. Otherwise there would be fewer far-right rallies in Berlin or Paris! You have to make the effort to awaken to the incredible opportunity for cheap, easy and fun travel that exists within your hometown. The first step is simply to become curious.

I grew up in Auckland, the biggest city in New Zealand. It's also the most international — almost 40 per cent of Aucklanders were born overseas. They include Pacific Islanders from places like Samoa and Tahiti, Europeans (including many British immigrants, like my parents) and Asians from China, Indonesia, the Philippines, India, Malaysia and so on. And yet, weirdly enough, Auckland isn't a haven of multicultural dynamism. There are neighbourhoods in Auckland where you would struggle to see a brown face, and other neighbourhoods where whites, or pākehā, are in the minority.

The school I went to is nestled between two very different neighbourhoods: affluent and largely white

St Heliers and less affluent and largely Asian and Pacific Island Glen Innes. As such, our school could have thrived as a microcosm of integration, diversity and acceptance, with each kid learning about their classmate's interesting culture and language. But it didn't quite happen that way. Don't get me wrong, things are generally pretty good and it's unlikely that many kids experienced overt racism. It's just that, when the classroom doors opened, kids seemed to naturally seek out others who looked like them to hang out with. Sure, there were Asians in my group, but they tended to be second or third generation; in other words they acted and sounded Kiwi — whatever that means. The first-generation Asians, including the exchange students, would segregate themselves off during lunchtime, always in the same classroom. I entered once. The only white face in there, I swear I remember chopsticks clattering onto the floor in shock as the denizens quit speaking in their varied foreign tongues and stared at this pale interloper. Perhaps this prepared me in some way for my later experience in China; soon enough I'd learn what it was like to be the interloper on campus.

I wish I'd had more curiosity back then. If I'd decided to learn Mandarin as a kid it would have saved me ongoing frustration and struggle in later life. Perhaps I'd have places to stay now in Bali, Manila or Kuala Lumpur. But while curiosity is a defining drive in us humans, so too is the need to surround ourselves with what seems comfortable and familiar. But you aren't a kid any more. There's no excuse not to make an effort to learn a little bit more about those foreigners who share your home town. You could search on Facebook or Meetup for language exchange gatherings. You could hang out, as I do, in Chinatown and make a fool of yourself by trying to order in Mandarin. The mechanics are very easy for anyone with an internet connection to figure out; what matters most is the attitude. And hopefully by doing so, you will start something of a trend.

When locals learn more about the new arrivals in their country and begin to humanize them, they are likely to dampen their reactionary hate with compassion or even affection. In fact, research cited in *The Independent* in 2014 showed that merely moving to a district with a more diverse population made white people statistically less racist.[4] They called this phenomenon 'passive tolerance', the idea that even passing exposure to people of different races makes people less afraid, more trusting and less hateful. But we are travellers and will not settle simply for *tolerance!* Travel at its best causes you to humanize those who look and seem different to you. And many of the benefits of travel can be achieved right in your hometown. So next time you return home from a trip, try to leave your brain in 'travel mode' for just a little bit longer.

TRAVEL PRACTICE
Be a traveller, even when you're at home

Perhaps the greatest thing that travel teaches us is to see the world as a series of adventures, with the people we meet upon the way not as strangers to be judged or feared, but interesting 'locals' to be learned about. The average resident of London has probably met dozens of people from the Indian Subcontinent over the course of their everyday life, but probably has never stopped to ask the searching questions about where, exactly, these people come from and what life is like over there. Take this same person, put them on a plane to Mumbai and plonk them down in a bar next to a local, and watch as the interesting conversation takes place.

Conversations with people from foreign lands are one of the purest joys of travel. When we are overseas, we turn on our curiosity switch. We hit pause on our prejudices. We don't get offended. We stop thinking of ourselves. We get lost in the moment, and in the story. This attitude of open-minded interest isn't magically generated by prolonged exposure to airline food! It is a persona we inhabit when we travel, when we imbue every interaction with a sense of the profound, and view every individual we encounter as an opportunity to learn. At home, we don't often do this. We encounter dozens of strangers every day, usually when we are in a hurry to get somewhere important. Learning the life story of everyone we encounter would be impossible. But we can still try, just a little.

Choose just one new person a day — or even one a week — and make an effort to learn something about them. Choose someone who comes from a different background to you and therefore likely has a different perspective on the world. Treat them as you would a local you were lucky enough to encounter in an exotic land. Indulge in the moment with curiosity. And see where it leads you.

Of course, at home as abroad, not everyone you meet will be a shining example of virtue and hospitality. There are people everywhere who, for the sake of your physical and mental health, should be avoided! But how do we, as travellers, learn who we should connect with, and who we should run from?

6.
The trust issue

It was hot in Istanbul. A little too hot. Sweat kept
dripping down into my eyes, and I paused on the
chaotic street to remove my glasses and wipe my face.
The tourist centre wasn't far away, but I was the only
clearly non-local on this particular street. And this
fact hadn't gone unnoticed.

As a foreigner in busy cities like Istanbul,
Marrakesh or Bangkok, popular with tourists, you
quickly get used to the touts. The tuk-tuks following
you down the street. The hawkers grabbing your arm.
The shoe shiners and trinket sellers, desperate for
your friendship. 'Hey, where are you from?' In Asia,
most are persistent but harmless. In North Africa, they
can be physically aggressive. In Istanbul, they can be
tricky, but they normally leave you alone if you look
sufficiently purposeful and refuse to engage.

But this guy was different. A short man of African
descent had been following me for some time, shouting
for some reason in French. '*Monsieur. Monsieur!
Monsieur!!!*' I kept my head down and ignored him, and
kept walking. It was my second time in the city and
while my friends were out sightseeing, my goal was to

find a nice local tea shop and catch up on reading. I had brought my Kindle along for this purpose, safely zipped inside my backpack. Or so I thought.

The short man had been gaining ground for 100 metres or so, and when I was forced to stop at an intersection, he finally caught me. He was waving something in my face. I went to turn away — a reflex borne of frequent exposure to such harassment in various countries and over the previous few days — when I recognized what he was holding. It was my Kindle! Not safe in my bag, as I had thought, but in the hands of this scruffy stranger.

He handed over my Kindle while I awkwardly thanked him. 'Many pickpockets here so. Very many. But you can always trust the black people.' His grin then became slightly imploring, and he began asking me where I was from, and making all the signs of someone who was about to become a very close personal friend unless I soon handed over a tip. So, politely yet firmly, I thanked the man for returning my device and told him I was late for meeting a friend. I ducked across the road, turned down a side street and lost him.

Knowing who to trust when travelling in some of the less tame corners of the world is a persistent conundrum for the intrepid traveller. Gullibly chat to every local who approaches you, and you will likely soon find yourself liberated of all of your possessions. Haughtily brush aside every approach, and you miss out on some genuinely well-meaning offers of friendship and local knowledge — fantastic opportunities to connect with a place on a real level. So how do you know if that sweaty stranger approaching you on the corner of a busy foreign street is friend, foe or something in between?

The persistent gentleman in Istanbul who followed me and returned my Kindle may have just been a good Samaritan helping out. Like an idiot, I had put the device in the outer pocket of my bag. Maybe it had fallen out, the man had seen it, picked it up and followed me, perhaps in the hopes of getting some money in return. Or maybe he had himself swiped it, hoping to find something of more value — and when he realized it wasn't a wallet but merely a very out-of-date reading device, he thought that by returning it he might get a greater tip.

Overthinking these little scenarios is a major part of the travel experience. Things don't work abroad the same way they do at home. The signals we generally look for to tell us who is trustworthy and who isn't don't always apply. And the stakes can be pretty high. If I had refused to trust every slightly scruffy looking stranger who said hi to me in a foreign country, I would have missed out on drinking rum and playing chess with locals in Havana, talking about football with Buddhist monks in Tibet, being invited to dinner with a local family at a border town in Turkmenistan and, well, drinking numerous local spirits in numerous dodgy places and hearing a whole bunch of great stories! Of course, had I trusted everyone who had approached me with a slightly too enthusiastic grin I'd probably have no kidneys, less money, and be writing this book from a Serbian prison cell (that one could almost have been true — maybe I'll tell ya sometime if you buy me a drink and promise to keep it to yourself).

The taxonomy of our perceived 'stranger danger' could be the subject of an entire book in itself. The writer Joe Aultman-Moore reflected on this in an Intrepid Times article about hitchhiking in Ireland. He describes happily hopping into a vehicle driven by a friendly local who warns him and his friend in sombre tones to avoid the 'gypsies' on the highway. 'Don't trust 'em. They'll rob you blind or worse.' After a hair-raising drive during which their new chauffeur managed to polish off about eight beers — hilariously described in Joe's article entitled 'Fear and Loathing on an Irish Highway' — the passengers stagger shakily out onto the side of the road. They sleep rough, and the next morning encounter a couple of young Traveller men — the very 'gypsies' they had been warned against. At first they feel fear, but then a friendly encounter ensues, with no one the least bit threatened. After this incident, Joe reflects:

[The oldest] was no more than fifteen but had fiery eyes and the stern hawk-face of an older man. I saw the mirror of who I might have been under different circumstances. I saw that we are afraid of those who are most nearly like ourselves and that the most dangerous people in the world are those that are a danger to themselves.

Knowing when to trust and when to run is critical for meaningful travel. I haven't always gotten it right, but over the years and through countless conversations with other travellers — young, old, male, female, black, white and everything in between — I have distilled a few tips and principles which should help. It begins with an obvious yet very important understanding:

A traveller is a resource

Most people with the means to travel for leisure are from reasonably wealthy countries. Even though you might feel you do not have two dimes to rub together, in most parts of the world the fact that you are a foreigner will cause people to assume that you are rich, or even if you're not, that you know someone who is — at least by local standards.

Seeing a traveller as a potential means of resource has its obvious manifestations: tourist trap restaurants, pickpockets, scammers. These tend to be easy enough to avoid with a bit of common sense (i.e. don't put your Kindle in the outer pocket of your backpack, and stay away from that restaurant with the big sign out the front saying 'authentic local cuisine' in English!)

Taxis are one situation in which avoiding the obvious

traps gets a little trickier. In a taxi, you are extremely vulnerable as the driver has you basically imprisoned. Normally, the worst that can happen to you is that you get overcharged by a few dollars, sometimes more. But there are stories out there about worse things happening.

When it comes to dodgy taxi drivers, rip-off businesses, crooked money changers and the like, you are dealing with people who are in the business of targeting tourists. They are likely to hang out at tourist areas, speak a bit of English, and aggressively solicit your business. Those who are in the more local parts of town and who are just minding their own business are less likely to have a prepared scam up their sleeve, so they'll probably just treat you like any other customer. At least, this has been my experience pretty much everywhere I have visited.

There is a more subtle expression of the 'traveller as a resource' mentality, however. This is when locals will approach you with totally innocent intention — they wouldn't dream of stealing your wallet and they don't even have anything to sell — but they will expect some benefit from you nonetheless. Mostly, this falls into the 'charming the first time, annoying the 500th time' category (i.e. people approach you wanting to practise their English, and take a few selfies to boost their social media cred). This is pretty harmless and can lead to more genuine interactions, and a chance for you to ask questions and get to know the culture a little better.

If you're staying at a place a bit longer, this more subtle 'resource' attitude can become a bit more problematic. While living in China, I was naively roped into lending my 'foreignness' to various locals who would benefit by showing potential customers their supposed international reach. You might find people ask you for help with immigrating to your home country or to give language lessons to their kid. Or to help them with the English in their thesis — a 'small' job that may end up taking half your week! While making these requests does not make your local friends evil, it shows the importance of setting boundaries. Be aware of the value that you will be perceived to bring as an outsider, and be willing to refuse to lend your presence, English skills, time or money if you don't wish to. Genuine local friends will understand.

You are interesting, too

Avarice is not the only motivation that locals will have for wanting to get to know you as a traveller. Far from it. In fact, when you travel outside of the main tourist hotspots, you'll discover that you are often just as interesting to the locals as they are to you! At a monastery in Mandalay in Myanmar, I was surprised to find monks coming up to me asking for selfies and quizzing me on life in New Zealand. This afforded a fantastic opportunity for me to also ask them about their lives. We were all loving it!

For those in countries like Myanmar who are too poor to travel abroad due to the weakness of their currency, or who are prevented from travelling abroad due to the policies of governments like that of Cuba, meeting foreign travellers is often their only opportunity to make friends from overseas. Whether you're in Tibet, Turkmenistan or Timbuktu, you can bet that most young people you meet will have grown up listening to Beyoncé, Eminem and Kanye West, and will know far more *How I Met Your Mother* quotes than you do, and will be absolutely thrilled to talk to you about it.

As with picking a taxi, those who are going to be genuinely interested in meeting you because they get the same kick out of it that you do from travel, are not likely to be hanging around airports or tourist sights lying in wait, ready to pounce on the first traveller they see with a selfie stick. They're more likely to be hanging out with their mates in bars outside the tourist district, and one of them may, showing off to his friends, approach your group and say hi. Provided you're in a public place and you feel safe, you might wish to continue the conversation. You can also search the internet for the local 'English corner' or 'English café', where you're bound to find a group of locals interested in chatting, or use an app like Couchsurfing to arrange for a local meeting ahead of time.

In my experience the best such meetings happen out of the blue; you just need to be ready for them. While sitting alone one evening at a small bar on the outskirts of Kyoto, the owner, who spoke no English, called her son to come and hang out with me. We ended up chatting all night comparing our experiences of life, music and study. While trying vainly to find a few essential supplies at a convenience store near the Turkmen–Uzbek border, a young man in a baseball cap stepped in as a translator — he'd just returned from a semester abroad in the United States, and took me and my friend to meet his mother who lived around the corner. These serendipitous occurrences are the lifeblood of meaningful travel. And if you travel long enough outside of the obvious spots they will happen to you.

> *After a day in the jungle with the elephants, our group returned to the village, muddy, sweaty and tired. I was nearing my homestay, when Grandma looked me up and down. My pants were torn, I had leech bites on my arms and legs, and my hand was wounded, wrapped and covered in mud. I gave Grandma a smile and nod, and wished I could tell her a joke. Instead, I continued to smile and went upstairs to my room.*
>
> *Later, after dinner, the sanctuary owner came over to me, face full of worry and asked me if I was okay and having a good time. The villagers were concerned about me; they thought I had hurt myself in the jungle. I laughed and assured her 'everything is perfect'. For the rest of my visit,*

> *Grandma and I would give each other a smile and laugh before I would leave for the jungle and I felt a beautiful sense of protection.*
>
> Tiffany P, United States

Considerations for female travellers

Women have concerns when travelling that men do not. Particularly in places like India and the Middle East where local attitudes towards women are less progressive, the attention foreign women receive can be over the top and extremely unpleasant, not to mention dangerous. There is also the added question in any interaction with a male local as to whether this man has a non-financial ulterior motive for making the approach.

Having met dozens of female travellers far more intrepid than I am who have ventured — often solo — through Morocco, Iran, Latin America and beyond, I can tell you with certainty that being born a woman does not have to stop you from having amazing local experiences and meeting friends all around the world.

A while ago a British woman called Rachael Rowe contacted me about a story she had written on the *Zurkhanehs*, or Houses of Strength, that are popular with locals in Iran. I asked her to write another article, focusing specifically on her experience as a western woman travelling alone in the conservative Islamic Republic of Iran.

In that article, Rachael recounts one episode when walking back to her accommodation after visiting some ornate Persian gardens:

> A party of schoolchildren were intrigued to see me walking through the gardens and ran up to me. In some countries there would be requests for pens but here in Iran each child offered me something from their lunch box — and insisted I try something. I didn't need to eat for the rest of the day!
>
> A car full of people screeched to a halt in front of me as I walked back to my hotel. What did they want? Two men jumped out and I prepared for the worst. But smiling, they simply welcomed me to Iran and thanked me for visiting the country, before driving away.

When it comes to staying safe, regardless of your gender, your own gut feeling will be the best guide of all. If invited by a local to dinner at their place, consider asking to bring a friend, and inviting along a strapping companion from your youth hostel. At the very least, let a friend know where you are going. Anyone who seems overly eager should be avoided. Understand the flavour of the city you are in. Unfortunately in tourist towns such as Bangkok in Thailand or Jaipur in India there's a very high likelihood that the locals who are bold enough to approach you have an eye on getting some money off of you, and we'll talk more about this unfortunate phenomenon in the next chapter. But if you're in a place with less tourist traffic, then there's a real possibility that friendliness has no ulterior motive — they may be just as innocently interested in you as you are in them, and the tradition of hospitality runs strong in much of the world.

TRAVEL PRACTICE
Talk to strangers, judiciously

It is hard to know who to trust when travelling, and every stranger in a strange land is vulnerable. There are some situations that should obviously be avoided, and certain behaviours like walking along at night in a bad part of town would be unwise anywhere in the world, even in your home town. The internet is not always our friend here: Google any particular city or country alongside the word 'safety' and you're likely to encounter forum threads full of horror stories of theft and harassment. It sounds trivial to say that bad things happen everywhere. And yes, you probably are more likely to suffer misfortune at the hands of strangers when you travel. You are also more likely to be stricken with food poisoning. Or to get lost. Or to meet interesting people. To gather stories that you will be telling for the rest of your life. To experience and *feel* beauty in a whole new way. To learn more about the world, and about yourself, in an hour than you would in an average year. Travel makes everything — good and bad —

more likely. So calculate the risk versus reward of every situation.

So much is lost if we close ourselves off to the opportunity of becoming friends with those strangers who we'll inevitably encounter far from home. This is not a call to let your guard down, nor is it an invitation to take foolish risks.

There are scammers and thugs the world over, but it pays to remember that not everyone you meet overseas is out to get you. Trust your intuition, and ask yourself what you think the intentions are of the person you're meeting, taking into account the circumstances in which you met them. And if things seem okay to you and you have a clear escape just in case things go wrong, then tentatively but openly move forward with the conversation. You may just have made a friend for life.

7.
The mass tourism issue

Many of us travel in order to change ourselves. But on the way, we seem to be changing much of the world around us. Fortunately for us, the problems of mass tourism are misunderstood, and surprisingly easy to mitigate.

Everything you see is for you. The locals greet you in English though it is not their native language, and attempt to entice you into their restaurant. Taxis follow you, their drivers announcing the names of famous tourist sights, begging you to enter. Herds of affluent travellers huddle, debate the merits of several adjacent Irish pubs, and finally sink morosely into chairs wiping the sweat from their face. Then they smile, take a selfie, and flip seriously through filters before returning to their hotel or hostel to freshen up before the next 'sight'.

I could be describing Bangkok. Or Venice. Or Prague. It doesn't matter. Travel changes the places that we visit — and not always for the better. Particularly in countries that are on average poorer than the United States or Western Europe, the lure of the tourist dollar is too much to resist. Fuck working in the field, in the factory, or getting a long and costly education. Just sell overpriced tours, hastily cooked noodles or mass-produced souvenirs to tourists. They can afford it. A few US dollars can feed your family for a week.

We travel to see the world as it is. But by doing so, we often force it to become something else. Tourism is a legitimate portion of the economies of many developed countries — not least of all New Zealand, Australia, the United States and Great Britain. But we as tourists can no longer, as the saying goes, 'take only pictures and leave only footprints'. Our mere presence has a warping effect on everywhere we visit. The more tourists come, the more a place changes. But the negative aspects of this have been exaggerated. Take a walk with me, please, down a noisy street in Bangkok, and I'll show you what I mean.

Local life is harder to kill than we fear

The hedonism of Bangkok's Khao San Road is intoxicating. At night, neighbouring bars blast top 40 tracks, one playing Rihanna, next door Flo Rida, as sweat-drenched locals sell beers out of coolers in the middle of the street. Other entrepreneurs approach you with spiders on a stick. There are stalls selling noodles and skewered meat. Some sell balloons filled, apparently, with nitrous oxide — tourists buy and inhale them for a cheap high. Locals come and thrust advertisements for the lewd 'ping pong shows' at you, containing vulgar descriptions of all of the things that women will perform, for a fee.

The street is a sight in itself — worth seeing, if only briefly. However, there is no doubt that Khao San Road, like so much of Bangkok, is an invention designed entirely to extract the tourist dollar. This is no secret. It is an artificial sight, far removed from any promise of 'authenticity'. Classic evidence of the 'destruction' caused by over-tourism. And yet even here, it is not too hard to slip away and discover yourself alone, far removed from hollering backpackers and obese sex-tourists, chatting over a beer with locals from Bangkok and immigrants from neighbouring Myanmar. In fact, all it took me, on my last visit, was a ten-minute walk, a few back alleys, a quiet bar outside and the willingness to say hello.

Finding the dry land of reality amid the overflowing waters of mass tourism is easier if you travel alone. Big groups create their own worlds. But the solo traveller is normally nimble enough to slip away. The signs aren't that hard to spot: English fades away from the shop names and restaurant menus, until all that's left is the local language. You see fewer and fewer tourists. Instead, locals appear to be living life for themselves. Couples sit and eat together. People queue outside the grocery shops. This is where you pause, take it in, and find a place to base yourself for a while. You may end up talking to someone. You may not. But, for a short while at least, you are the only interloper in this city. Enjoy the show.

119

Culture is resilient. It has survived for thousands of years. Yes, it will change and adapt and utilize the resources that are offered to it. And yes, you are one of these resources. But that doesn't mean it is no longer real. What many of us see as a place being ruined by over-tourism and the related harms may not in fact be a fair assessment of reality.

The real tourist trap is in your mind
For all the legitimate harm that is created by over-tourism, and the much-feared homogenizing effect of globalization (or 'globalism' as some politicians have taken to calling it), it takes very little effort when overseas to discover the truth: local life isn't that easy to kill. Just because the people of Bangkok enjoy modern transportation, iPhones and McDonald's does not mean that Thai culture is dead. Tourists have colonized much of Bangkok, sure, but not all of it. But things seem worse than they are. There are two reasons for this.

The first reason why it is so easy to throw up our hands, scream that the sky is falling down and mass tourism is ruining the world, is that the tourist industry is determined not to let you stray off the beaten path. The ads on your social media, the results on your next Google search, even the recommendations you receive from friends, are all the result of a mass travel industrial complex that is designed to concentrate tourists into the one area. Your hotel gets your dollar because it is near the famous sights you have read about. Restaurants and souvenir shops sprout up near these sights, so everything you need is in one place. No need to go further. For many travellers, this is all they will see. This is their loss. It's also your gain: because most tourists are so willing to be corralled into their little bubble, it means this bubble is easy to escape. Just walk for ten minutes, away from the signs, and you'll find yourself somewhere real.

The second reason is a matter of perception. There is a certain romanticism, perpetuated by tour companies, American movies and our own narcissism, in imagining that people in distant lands would be living in primitive paradise were it not for our 'toxic' (we say with a self-deprecating smile) influence. Thailand should be a land of humble rice farmers. Venice a painting of gondoliers slipping lazily down canals. Should a country deviate from our romantic perception, then we must have broken it. Poor them and shame on us. When brought to the surface, the arrogance and absurdity of this perspective is easy to see through. But to what extent do we live with a level of this attitude, lying unexamined on the periphery of our consciousness?

While the first problem is a practical matter that can easily be solved by changing your travel *tactics*, the second is a matter of perception that can only be solved by improving your *understanding*. One of the main purposes of this book is to arm you with the tools to do just that. To help you observe without prejudice and breathe in the scenery before you without judgment. A descendant of Vietnamese farmers has just as much right to an iPhone and a cheeseburger as you have. This doesn't mean that he stops being Vietnamese or that he becomes a poor clone of the United States. The development of a culture does not equal its death. This shift in perception will not only help you allay traveller's guilt; it will allow you to breathe in a more real and honest version of each place you visit. The mission, after all, is to travel and see the reality of a place, not the shadow of what you or your guidebook thinks it ought to be. But this reality can be harder to pin down than it seems ...

Everyone I knew who had visited Thailand had told me how incredibly beautiful Chiang Mai was, so I was very disappointed to find, during my first taxi ride through the city, that the streets were lined with worn-down buildings, the air was heavy with motorbike fumes, and the traditional Thai architecture I had been waiting for was nowhere to be seen. It took me several months to discover that most of Chiang Mai's beauty was hidden behind the paint-chipped walls and cheap plastic signage. Its history was buried in attempts at modernity, and I often had to go through a back alley to find the most beautiful temples, enter a sketchy storefront to find the best cafés, and be in the right place at the right time to see the expansive street markets pop up out of nowhere.

> *Chiang Mai taught me not to trust first impressions when travelling, that beauty may take time to make itself known.*
>
> Jennifer R, United States

To see a place is to change it; but it's okay to lose your certainty

On 5 December 1901, a man called Werner was born in the German city of Wurzburg. At the tender age of 26, young Werner developed his uncertainty principle, which he first referred to as *ungenauigkeit*, which could be translated as imprecision. Five years later, he was awarded the Nobel Prize in Physics for 'the creation of quantum mechanics'. Well, I may be approaching the age at which Werner Heisenberg won his Nobel, and all I can claim to have created is a moderately popular travel website and a couple of lethal twists on the classic martini recipe. Bloody overachiever.

The Heisenberg uncertainty principle is quite complex. I don't pretend to fully understand it in the context of physics: it has something to do with the fact that objects have properties of both waves and particles. Particles can only be in one place at a time. Waves can be in lots of places at a time. So if we say 'you're a particle, therefore you're here' we can't measure its wave properties. And if we say 'you're a wave, so you're all over the place' we can't measure it as a particle. So in order to measure the object across both categorizations we have to relax our measurements. We have to lose our precision and abandon our certainty. In order to know something completely, in all of its aspects, we cannot know it accurately.

The Heisenberg uncertainty principle is often confused by non-physics people (i.e. me) with the 'observer effect.' The observer effect states that observing a phenomenon itself is often enough to change this phenomenon. This could be because of the equipment you use, the way the experiment is structured, or just because the universe is having a bad day and wants to piss you off. You cannot know something without observing it, but by observing it you are changing it, so you can never truly know what it was like before you took a peek, you nosy physicist you. The observer effect played an important role in how Heisenberg developed his uncertainty principle. As did a bloke called Planck and his mate Constant. Stupid name, if you ask me. Where was I?

The connection between the observer effect and travel is clear: merely by visiting a foreign country and observing it, you are changing it. The more people who visit and observe, the more it changes. The more tourists visit a place, the greater the transformative effect they have on the place they visit. But even when travelling solo, you will have an impact on where you visit. As we have explored earlier, as a foreign visitor in a less well-trodden destination you are as interesting, foreign, exotic and exciting to the locals as they are to you. They will approach you. Chat to you. Host you. Take photos of you. Change their behaviour because of you. You can try to get out of the way, but the power of your presence will be felt, regardless.

This leads us to what I'll call the Heisenberg travel principle. In order to know a place as a foreigner, you cannot know it as a local. And in order to know it as a local, you cannot know it as a tourist. Locals linger, economize, speak the language, blend in. They don't look up at the sights or snap photos. They aren't focused on the experience of *being there*; they are busy living their lives. Tourists pass through quickly. Locals relish the moment. They indulge. They stop and smell the roses. In order to travel and have both — as I have tried to do on my recent travels — you have to, to some extent, lose your commitment to either. As an expat in Budapest, where I am writing this book from, I live in some ways like a local: the city is the backdrop to my day-to-day life. But I also live like a tourist. I make an effort to explore, to see, to discover and to learn. I am not as integrated as a local — far from it — nor am I as casual and fleeting as a tourist. In order to try to have the best of both worlds, I fail at both.

My husband and I decided to take only scenic routes on our recent trip in Victoria, Australia. Our drive from Rawson to Melbourne Airport, which normally takes two hours via tolled motor-ways, took us six hours ... but we experienced

125

gigantic 250-foot mountain ash trees in deep fern valleys, apple orchards in lush vineyards, small townships full of quirky stuff ... what a great way to end our holiday!

Joan G, Singapore

Imprecision, uncertainty and the observer effect can be mitigated by patience, by making an effort to learn the language, by trying to blend in, but they cannot fully be eliminated. Like a physicist calculating the location and velocity of something that can be both wave and particle, and trying to reconcile these measurements in a neat 'quantum object', the uncertainty you witness and the changes caused by the mere fact that you're observing are something you have to account for in your calculations. Merely by being conscious of them, you can enhance the richness and experience of travel, as you humbly factor the impact of your presence into the equation.

This loss of certainty is doubtless a headache for physicists, but for us travellers it can actually be a blessing. The world stubbornly refuses to be pinned down or neatly categorized. It demands constant reevaluation, forcing the right-minded traveller to continually question their beliefs and assumptions. This is a good thing.

Okay, enough science and maths. Let's come back down to earth. Given the uncertainties we have explored, how can you as a traveller behave responsibly, so you form meaningful connections with the right kind of people, and make the inevitable impact of your presence a positive rather than a negative thing? As with so much in this life, it begins with your wallet.

TRAVEL PRACTICE
Spend your money where others don't

Before we got onto this lovely metaphorical physics tangent, we looked at how the tourist bubble is actually pretty easy to avoid. Don't book the same tours and stay in the same hotels as everyone else. Or if you do, wander off the beaten track a bit. Head away from the sights, get off the mini map in your guidebook, and find yourself in the part of the city or town where real life occurs. It can help, not only as an altruistic investment but also as a way of enhancing your own travel experience, to spend your money in the same way.

Next time you're away from home and you're thinking about spending some coin — be it on a tour, a hotel room, a sandwich or a beer — first ask yourself *who* that money is really going to. It is unlikely to be the friendly local to whom you're handing over your credit card. Your cute boutique hotel may well be owned by an American or British chain, or by a corrupt local government. The internet can be a useful resource for finding companies authentically owned by locals. Instead of big hotels, look for homestays. Even Airbnb is a step up — at least your money is going directly to a local who may end up becoming your friend and guide, and offering insight into the local community.

When speaking to locals, don't ask them where you should go as a tourist. In the name of being helpful, they'll send you to the same restaurants, artificially packaged cultural shows and tourist traps that everyone goes to. It's not that they think you're stupid; it's just genuinely what the evidence before their eyes tells them that westerners like. Instead, ask them where *they* go. Offer to take them out for a bite.

As a traveller it's basically impossible not to participate to some degree with the tourist economy, and some percentage of what you spend anywhere will always be skimmed away by governments and large corporations. But at least by spending your money consciously in places where the other tourists don't, you're helping to spread the love and keep something real alive.

Beyond the issue of spending, there's the matter of the kind of person you are as you travel. It's easy to snap a selfie with a local. It's harder, but ultimately more rewarding, to listen, to give them an audience, to make them feel important as a representative of their people and culture. We explored earlier how putting a place in some sort of context can help you arm yourself with the right kind of questions to ask. Apply this. This alone makes you part of the solution to ignorance, rather than part of the mass tourism problem. You express curiosity, which results in your own enlightenment and education. And by expressing curiosity and listening with genuine interest, you show a different side of tourism to the locals you are speaking to.

8.
The climate issue

The problems of mass tourism can be misunderstood and are easier to mitigate than most fear. However, travel done carelessly, recklessly and without good sense can cause real harm — not least of all to the environment.

The climate activist Greta Thunberg, who shot to global stardom in 2019, is a highlight at climate conferences around the world, but she refuses to fly because of the detrimental effect air travel has on the environment. So public is her refusal to fly that a BBC journalist was ridiculed in the global press for taking a flight in order to interview her. This is a tricky conundrum for travellers. Of course, we love planet Earth. We love it so much that we want to see it all, understand it in all its various contrasts, diversity and complexity. The best way to do this is often to fly — the convenience and affordability of air travel is just too good for many of us to pass up. But if these flights are ravaging the planet that we love, making the very act of seeing the world an act of destruction, should we travel less? Or restrict ourselves to places a bus or train ride away?

This is a serious question — saying no to flying drastically shrinks the amount of the world realistically available for most of us to see. So this problem deserves some analysis. To help get clarity with the numbers, let's take the exceptional case of one Tom Stuker, a man who got press attention for clocking about 29 million so-called 'butt-in-seat' kilometres (18 million miles) on United Airlines (i.e. the distance he actually flew). An article in the *New York Times* on the damage that the explosion of cheap and accessible flights is doing to the environment estimates that one passenger's share of the emissions caused by a 4025-kilometre (2500-mile) flight is '35 square feet of Arctic summer sea ice'.[1] If we are to take the *New York Times'* maths and apply it to Tom, then our jetsetting friend's total share of Arctic summer sea ice destruction would be around 2.3 hectares (252,000 square feet).

While writing this chapter, I was expecting to do some maths and show you that Tom's trail of destruction across a lifetime of 29 million kilometres of travel would be 'the equivalent of Iceland' or something shocking like that. The area of Iceland, a tiny country, is a little over 50,000 square kilometres (34,000 square miles). When converted from square feet, Tom's alleged 'lifetime of destruction' works out to 0.0090392562 square miles of sea ice destruction. That's ... well. Sure, it isn't *that* much. But for just one person, multiplied by millions of flyers per year, it does add up. So, can this harm be mitigated?

The article discusses and dismisses some alternatives for international travel: cruise ships (worse polluters than planes), and the carbon offsets you can sometimes buy when booking your flight. These are described as a mechanism used to assuage guilt that don't actually guarantee any long-term benefits to the environment — the money will most likely end up in the hands of some giant company who, after dutifully using it to plant a bunch of trees, may just chop the lot down next year and burn them.

When researching this topic, the advice out there was contradictory. Some advocated a reduction or even cessation of flying. But that's extreme, uncalled for and misguided, as we will see below. Others suggested taking only direct flights, rather than longer journeys interrupted by a stop. Of course, travellers only take longer routes to avoid drastically higher fares. Perhaps the airline industry itself could prioritize efficiency over profit. But that will never happen without regulation.

In the end, we humble travellers are left to take the blame for an environmental catastrophe of which we are a small and relatively powerless part. While we as individuals can and should be conscious of the environmental impact of our decisions, focusing so heavily on the culpability of a single airline passenger takes the important environmental conversation dangerously off focus.

Is air travel a red herring?

In the big picture of climate change, how much of a sinner really is the travel industry? A January 2020 *Forbes* article on the flight shaming phenomenon explains that the commercial aviation industry accounts for just 2 per cent of emissions globally.[2] Compare this with the fashion industry, which accounts for 8 per cent of global emissions, and the food industry, which accounts for 25 per cent, and you may start to wonder … have flights been unfairly singled out? To put it another way, do you find yourself spending four times as much time thinking about the environmental impact of that new pair of shoes as you do for that cheap flight ticket? If you do, then I'd posit that you're quite unusual. There is so much noise in the anti-flying or so-called 'flight shaming' world now that many passionate travellers are seriously reflecting on whether or not flying around the world is worth it. Spoiler alert: It is.

Certainly, we need to think about the environmental impact of how we travel — but flying is not even the most damaging part of the *transport* industry … not even close.

The National Academic Press reports that only 17 per cent of the petroleum used by the transport industry is used in so-called 'non-highway transportation'.[3] This amount includes air, as well as rail, and oil pipelines. When you look at it in context like this, it seems a bit odd that the airline industry — which makes up a tiny fraction of the overall harm that humans are doing to the environment — has been singled out for so much criticism. And you know what, there's probably a reason for this, and you're not gonna like it.

In travel as well as politics, it pays to pause and ask yourself what is *not* being discussed. All of this discussion about the harm of flying, a minor accomplice at best in the rogue's gallery of climate harm, takes our attention away from the real criminals that are trashing our planet and cashing in their tax cuts in the process. The implied message here is simple: 'Citizen, please spend less time studying our dubious environmental policies and the tax cuts we still give to the petroleum industry, and more time wallowing in self-inflicted guilt over your next Ryanair flight. Don't ask about what we're doing to regulate the airlines. Ask yourself whether you really should be taking that bargain flight for your vacation. Maybe stay home, drive to the mall, and spend your money in my state instead.'

Outsourcing the responsibility for issues like climate change from governments and corporations to individual people is a well-established neoliberal political tactic. As Ronald Purser writes in *The Guardian*, 'The neoliberal order has imposed itself by stealth in the past few decades, widening inequality in pursuit of corporate wealth. People are expected to adapt to what this model demands of them. Stress has been pathologised and privatised, and the burden of managing it outsourced to individuals.'[4] 'It's not corrupt governments, entrenched

135

political interests and greedy corporations who are ruining the planet,' this attitude cries. 'It's you!' When brought to the surface, it's not hard to see the flaw in this logic. We would be better stewards of the planet if we voted consciously on this issue, rather than personally absorbing all of the guilt for a phenomenon that is largely beyond our individual power.

Why does this matter? Surely, flying less would be a good thing anyway, right? We can all do our bit. Well yes, and we can and should all do our bit to be more responsible consumers of fossil fuels. But when we start to see ourselves as the villains, and not the giants of industry, we lose sight of the real battle here. If we fly less, we potentially restrict our understanding of the world and quality of life; the ultimate difference we make to the environment is limited. If we travel, we increase our understanding of the world, and can use that increased understanding, sensitivity, influence and network to lobby and effect real change on the level that matters. Then our impact can truly be positive and profound.

There's another flaw behind the attitude of the 'don't fly, save the planet' argument. The logic underpinning the injunction on flying for the sake of the environment seems to assume that if you did not travel overseas, you would stay home, cycle everywhere, eat vegan meals and practise yoga. For some of you that is no doubt true, but for most of us it isn't. At home we drive. We shop. We participate in polluting industrial economies. While overseas we hike. We take trains and buses. We walk the city streets and judiciously spend our money on tours and businesses who share our values. Travel makes us better people. A world full of people who refuse to fly is not a better world.

My trip to an Italian city ended in serenity on the dunes of one of its beaches. I smelt the brackish sea while watching a boundless horizon of shining water. People ran along the sand, played with their kids, making my view of the Adriatic Sea even more attractive. Then I went down to the shore and a plastic horde surrounded my ankles. How could we have turned off our sense of beauty and ruined this seaside? I thought.

It was only then I realized that all those people weren't running or playing in the sand. They were collecting every small piece of plastic, leaving the beach better for future generations. As I joined them, I realized that humans had not turned off their sense of beauty. The feeling to create a better world was still alive.

Alessandro C, Italy

On a personal level, there are things you can and should do to mitigate to some extent the environmental costs of your journeys. Try to carpool. Take trains where possible, which have a much lower emission rate per person than other vehicles. If you drive, get a car with a good environmental rating. All of this makes a difference. But choosing not to take that flight, not to see the world, not to expose yourself to something new, is to put on your shoulders the blame for a catastrophe that ultimately is the fault of mass industry that can be better influenced by exercising your vote. Force politicians who actually are in a position to do something to finally take responsibility. And get out there, see the world, learn more about it, so you can make informed decisions about who you elect and why.

I'm not saying there's no cost to travel. There is. But in perspective, it's less than we may be persuaded to think, and it is far outweighed by the benefits discussed extensively elsewhere in this book.

The right to travel

There's a slightly uglier side to the environmental debate on travel that lurks not so much in article headlines, but in the comments section, in bar-room conversation, and to some guilty degree in the minds, particularly of younger, millennial travellers. 'Travel is a privilege, not a right,' this voice says. 'Humans are meant to stay in one place. Travelling this much is unnatural.' But when we were all forced to stay at home, were we really happier? Did we feel 'grounded', or did our frustrations hint at a deep-rooted need that we were not, at that time, able to meet?

Of course, modern technology has increased the

speed and distance we can travel within a lifetime, but the need to travel is deep-seated and human. We live in an age that gives us unique potential to connect with others from around the world. It is absurd to expect that we would neglect this privilege. Travel is, and has always been, an innate part of being human. No one understood this better than the late English travel writer Bruce Chatwin. In his travels, he set out to understand the issue of restlessness, trying to get to the heart of the question of why, even without practical necessity, throughout all of history and across the world there have always been those groups who have chosen to wander rather than to stay still. This quest to understand nomadic people from the Indigenous people of Australia (see his book *The Songlines*) to the nomadic tribes of Afghanistan and far beyond.

In a letter to his publisher, when discussing a concept for a book entitled *An Anatomy of Restlessness* (a book that he never wrote, but this would end up being the title for a wonderful collection of his posthumously published shorter works), Chatwin laid out in brief the broad strokes of his argument that the need to travel is part of being human:

> ... in becoming human, man had acquired, together with his straight legs and striding walk, a migratory 'drive' or instinct to walk long distances through the seasons; that this 'drive' was inseparable from his central nervous system; and, that, when warped in

conditions of settlement, it found outlets in violence, greed, status-seeking or a mania for the new.

Chatwin would compare, unfavourably, those ancient peoples who stayed still versus those whose culture had a nomadic bent. Writing movingly in *The Songlines*, he decries those 'sluggish and sedentary peoples, such as the Ancient Egyptians — with their concept of an afterlife journey through the Field of Reeds — [who] project on to the next world the journeys they failed to make in this one.'

To travel on foot, by train, on horseback, atop a camel, via budget airline or intergalactic spaceflight is, has and always will be a part of our nature as humans. It improves us not only through what it literally exposes us to, but also by fulfilling an inner need. Travel is justified itself in its own right. If you truly feel the need to travel, the only way to make peace with it is to heed the call. That ringing in your ears will not go away. And like it or not, we live in the age of aviation. No generation in history has ever been asked to turn their backs on the opportunities of their lifetime. And nor should we.

TRAVEL PRACTICE
Don't feel guilty about seeing the world
For the conscientious wanderer, guilt and travel have become inevitable companions. My challenge for you now is to release that guilt. Think about the reasons you might feel that travelling could be a harmful act.

And then consider the arguments in this chapter and the one in the previous chapter.

Here are a few of the most common arguments against travel, and a reminder of the answers to them.

- 'Mass tourism changes the local way of life.' This is often true, but that change is not always negative. It is wrong to romanticize poverty. And in cases where mass tourism is harmful, it is easy not to become part of the problem. Get off the beaten track and spend your money where others don't.

- 'Travel damages the environment.' Air travel makes up a tiny percentage of the petroleum used by the transport industry. The recent hype about the harms caused by flying is a red herring. It's a distraction from corporations and government policies that are really trashing the planet. Take the flight, and spend the time and energy you saved examining your local politicians' policies on the mining industry, chemical manufacturing, and their taxing of petroleum companies. Let this inform your vote and discourse. This will have a far greater positive effect on the world.

- 'It is more moral to stay at home.' Your home is not your prison. Seeing the world can help cure prejudice, ignorance and the far greater sin of being boring. If you feel the call to travel, it is your duty to heed it to the best of your abilities.

The next time you are challenged not to travel, to stay home, to see less of the world, remember these arguments.

9.
Blending in

It's early morning in a foreign city. Your alarm rings. You blearily stumble out of bed, shower, don your cleanest 'business casual' outfit and head into the elevator. Outside it's light already, and while this new world is still exciting to you, each day the novelty dims a bit as you find yourself lost in the crowd. The metro or tram is no longer scary. You have memorized the route and have a local transport card. Like everyone else around you, you jostle for space and robotically face the wall as you flash by stations filled with equally drowsy commuters.

You arrive at the station and emerge onto the street. Immediately, you notice that the homeless bloke, who is normally trying to break the world record for most cigarettes inhaled at once, is not at his usual corner. You worry; did Smokey Steve take his last puff? The queue at the local pancake stand is longer than usual. You take note. Do they have something special on today? You make it to the office or classroom, and sit down among colleagues or classmates from across the world, locals and foreigners alike. The day's work begins. In the evening you might join the mixed group for a beer, or you may once again blend in with the herd of commuters as you make your way back to your apartment.

You're far from home, but you're becoming a part of the fabric. So the question is, are you really a tourist, or even a traveller?

The idea most of us hold about travel is a pretty straightforward equation: person from country A leaves said country and visits country B. In country B, this person sometimes doesn't speak the language, knows few people, is equal parts confused and fascinated by this strange world. And then this person leaves, and heads back home or onto the next country.

But what if you stay?

It's an uncomfortable truism of our language that people from wealthy countries who move to poorer countries for economic opportunity are known as expats, while people from poorer countries who move to wealthier countries for the same reason are known as migrants, a politically loaded term. Migrants are expected to integrate on pain of derision and deportation. Expats are expected to stand out. They can happily live in a foreign country for years on end without learning a line of the local language, flashing in air-conditioned cars from gilded compound to gilded office and back again.

But there are those who try to fit in, at least in small ways. They study the local language, at least enough to empower them to eat at local hole-in-the-wall restaurants, not just at McDonald's or the Hilton. They live not in an exclusive walled compound but among the locals in an apartment block where they have to contend with grumpy neighbours and surly landlords. They commute cheek by jowl with locals day in, day out, recognizing the old lady with sharp elbows who would — and probably has — committed murder to get the last spare seat. They are still and always will be foreign. But they know the streets perhaps as well or better than the streets where they grew up. They mingle with locals and foreigners alike. They experience local bureaucracy, a frustrating and inescapable window into how a country lives.

Some expats refuse to 'travel' — despite being far from home, they cloak themselves in all the trappings of the familiar so they need not be threatened by the exotic. Others take a deep dive, becoming obsessive about the foreign culture, looking with embarrassed scorn on other foreigners, dating and perhaps marrying locals, and speaking more frequently in the local language than their mother tongue. Most, including myself during my brief periods of expatriation in China, Poland and Spain, are somewhere in the middle. But even for us middle-road expats there are magic moments, perhaps once a week at best if you're lucky, when you're ordering at your local hole-in-the-wall in the local language, where they know your name, and for a shining second you're the only foreigner in sight, but you don't feel foreign, you feel like you're part of the fabric of this new world.

For this fleeting moment you have broken the bubble. You have shed the perspective of where you come from. And you have become a *part* of where you are. Of course, then a herd of British tourists walks into your local: 'I've seen this place in *Lonely Planet*,' says one, beers are ordered, and the spell is broken. But for a second … just for a second … you had it. And once you have it, you'll never stop searching.

Check your privilege

While driving across Central Asia in a broken car with a group of mates from back home, a certain phrase would surface whenever one of us made a particularly daft or tone-deaf comment about the state of local life. 'Check your privilege.' Said in jest, the phrase doubtless carried with it the weight of a lot of meaning. It was an injunction to shed the perspective and expectations of home, and see the world we were passing through for what it was, not what we thought that it should be. It also masked a lingering feeling of guilt that we should be able to leave our stable jobs in our stable countries and roar across the world for a few months, passing by people far poorer than we, who, of course, never had and would likely never get such an opportunity.

The 'traveller's guilt' phenomenon is very real and can be an actual and also psychological barrier to properly experiencing your travels and connecting with those who you meet along the way. The world we live in is fundamentally unjust, as anyone who watches the news will see. In a world where the nation-state system keeps many people bound in one place and prevents many from pursuing economic opportunity, the rich will always protect their wealth and imbalance will remain. While being realistic about this, it's also important not to get carried away.

In Hans Rosling's thought-provoking book, *Factfulness: Ten reasons we're wrong about the world — and why things are better than you think*, the Swedish statistician shows how the psychological division we have in our minds of rich or 'developed' countries versus poor or 'developing' countries is outdated. It shows how media bias has conditioned us to think that life in poor countries is a daily struggle against death and disease. In fact, the world is better than what 80 per cent of us think it is. Despite the extent of recent challenges, the big-picture trends remain broadly positive. Poverty is decreasing across the board, and access to education and healthcare is rising. For example, most people think that only 20 per cent of girls in poor countries finish primary school. The answer is actually closer to 60 per cent. Most people also believe that the majority of the human population lives in poor countries (which are not quite as bad as we believe anyway). In fact, most people live in middle-income countries.[1]

Bill Gates said of Rosling's book, 'One of the most important books I've ever read, but don't let that put you off.' It deserves to be digested and appreciated properly. It is not a call for complacency. Of course, even 60 per cent of girls finishing primary school is far too low. But the numbers are getting better year on year. The world *is* improving. Not only that, but the types of lives that people lead are, around the world, more similar than we might expect. Rather than nationality or ethnicity or anything like that, the number one determinant of how someone lives is their income level. A person of low income in China will live a similar life in terms of hours spent at work, variety of food on offer, education level and so on, to a person of low income in Columbia. A rich American actually has more in common in terms of lifestyle with a rich Russian or Turkish person than they have with a poor American from their own city.

The smaller differences you need to blend in to see

Hans Rosling's explanation of the similarity of how we live our lives around the world based on our relative income level is one of the reasons why blending in as a traveller is actually easier and also more important than many may think. The scenario we painted on p. 143, of the expat at the metro station, probably doesn't feel that different to your commute at home. When you blend in enough to stop being blinded by the surface level differences — the strange language, exotic food, different looking people — you can start to understand the deep and profound similarities. And then, you can look for the more subtle distinctions that may reveal interesting stories and provide deeper insight into cultures and people.

For example, while riding the elevator every day from my apartment block in China, I eventually noticed that there was no fourth floor. This is a common feature of large buildings in China — the Chinese word for four, *si*, sounds like the word for death, so four is unlucky. Recently I was riding an overnight bus from Croatia to Hungary. While trying to find my seat, I noticed that the numbers skipped from twelve to fourteen. Row thirteen is often omitted on buses and airplanes in this part of the world. Even Lufthansa skips both thirteen and seventeen, so people who are superstitious about these numbers don't have to sit there! While in China, this superstition points to a fascinating cultural obsession with the power of homophones (eight, *ba*, is lucky because it sounds like *fa*, meaning to get rich), in Europe it connects us with our Classical roots. Seventeen in Roman numerals is XVII. The numerals can be rearranged to make the word *vixi* which can be translated as 'my life is over' in Latin. Number thirteen is associated with Jesus Christ and the number of people supposedly at the last supper. So there you go.

A bit silly though this example is, it shows how noticing the subtle, smaller things that you would probably only connect with once you'd 'settled in' to a place can actually lead you down a rabbit hole of cultural insight, possibly tracing back to a place's ancient history. Hans Rosling shows us that the world is not so strange, poor or different than we may feel it is. We travellers shouldn't let our guilt blind us to the immediate reality that we see. We should allow the rhythms of a place to wash over us, and then we'll notice what the more casual visitors do not.

The real bubble is financial, not geographic

What if you don't just want to experience middle-class life around the world? What if you want to see how life is lived at lower income levels, in order to better furnish yourself with an understanding of the world as it is for everyone in it? This is a noble objective if it is carried out a) from a place of pure curiosity and not the crass urge to gape at the poor, and b) if it is undertaken from a place of educated understanding that sheds exaggeration, sentimentality, or any sort of guilt-fuelled messianic complex.

'Poverty tourism' is the trend of companies taking rich tourists to look pityingly on *favelas* and ghettos. Those in lower income levels who live off the land as farmers or eke out a living on the margins of society in cities are not a remote alien race to be stared at as if in a zoo. They are all around us, everywhere you go, and in my experience are normally willing to say 'hi'.

In China, I would chat to labourers in my dodgy Chinese at cheap canteens around my university, people who lived lives very much at the bottom of the income chain. I was able to do so because, as we discussed in Chapter 7, it's easy to break out of the bubble and find yourself where the locals are. And once you make one connection, like a chain, you will find yourself being faced with opportunities to connect with people from all walks of life. A chance invitation from a local friend led to a stay in an isolated Sichuanese village where I was the only white face a small boy who lived there had ever seen.

Or take the case of Michael Falcone, a lawyer who gave up commercial life to live among the Sherpas in a remote Nepalese village. Seeking respite from the world, he asked a guiding agency in a touristic part of Kathmandu to help find him a hut or even a cave somewhere far away so he could meditate. The agency knew just the place. Soon, Michael found himself living at the foot of Mount Gaurishankar in a small Sherpa village. Despite being the only foreigner, not to mention the only westerner or lawyer in town, he soon found himself becoming part of local life.

> Within the month I felt like a member of the village. The nearest neighbours across the creek ran a small shop out of one room of their home, but every time I had made the trek, I had purchased nothing. Despite my being a foreigner, they showed no interest in selling me anything. Normally, I was simply invited into the kitchen area and served either a rice and vegetable stir-fry, or perhaps whole potatoes baked in the smoldering coals of an open fire with spicy chutney.

You too will and may have had similar experiences. You just need to show up, start the first conversation and be willing to say yes to invitations that might be a bit out of your comfort zone (for the safety side of this, see Chapter 6, 'The trust issue'). And of course, this process holds true back home as well as overseas. Many of us have just failed to think about it this way — our perceptions of this whole topic are deeply clouded.

There is something deeply uncomfortable about the income-level conversation and it feels somewhat grotesque to categorize people based on wealth. For many westerners there is a cultural taboo against discussing such matters, which makes even thinking about this in a clear-eyed manner difficult. We have also been so drenched in media narratives that either sensationalize the horrors of poverty, or romanticize a so-called simple, pure and minimalist lifestyle that it has now become perversely trendy for rich Silicon Valley types to want to emulate. This narrative becomes warped to the point where we think that the poor are alien and inaccessible unless we are aid workers, and that, perhaps more damagingly, this silent seething mass of the downtrodden makes up that majority of the world's population and basically the entire population of so-called 'developing' countries.

The nature of our understanding of how people actually live their lives abroad reared its head when photos emerged during the 2015 refugee crisis showing that many of the 'migrants' had smartphones. To some, this was 'proof' that these people were not real refugees at all. But what image do these pundits have, I wonder, of how people in Syria or Tunisia actually live their lives? Do they think electricity flows only in London and Paris, and the rest of the world is a dark, dangerous impoverished place where only the wealthy elite have smartphones? Poor Americans have them, so why shouldn't poor Syrians?

TRAVEL PRACTICE
Break your income bubble
The life you lead is as much a factor of your income level as it is where you are from. No matter how open minded and non-judgmental you are, it is just a fact of life that you will tend overall to hang out with people who are like you. The idea of the 'social media bubble' has been hashed out over and over again ... but have you ever thought about your *income* bubble?

In a lot of developed countries, the social separation of people based on their relative wealth happens from the moment you go to school. This follows you to university, and is of course represented in the connections you make at work. Even if you don't have a snobbish bone in your body, circumstances have likely engineered things in such a way that most people you associate with have similar levels of wealth to you. So break this bubble.

If you're a middle-class professional, branch out and start frequenting your local working-class watering hole. Hang out in some of the rougher parts of town. Make friends. Invite people over to your place and wait until the invitation is reciprocated. You may feel like you are travelling. You probably dress different to those you'll meet. You probably use a different vernacular, maybe even have a different accent — in the United Kingdom, accent is still closely associated with class and wealth, and this is also true of parts of the United States.

This is not a patronizing exercise in 'slumming it' — it's a conscious exercise in expanding your world. Many of the most interesting people I have met are just as comfortable drinking pints in a workmans' bar as they are sipping martinis on the roof of the Hyatt. Make the effort to expand your circle, and gradually — without acting or faking it — you will develop the knack of blending in with different groups and demographics. And when you do, you'll start to notice more, be let further in, and greatly enhance your understanding of the world.

Making an effort to branch out and meet people from different backgrounds, who live life in a different way, can happen at home as well as abroad. You need to be clear eyed about the reality of the world and willing to open conversations in unfamiliar settings, and say yes to the invitations that can, and inevitably will, follow. This will not only have immediate benefits at home, but it will make you a better traveller. It will improve your ability to blend in and feel comfortable wherever you are. And when you feel comfortable and make others feel comfortable around you, once your eyes have adjusted and you are no longer dazzled by the surface distinctions, you will notice more subtle points of contrast that could lead you to discover more stories and insights than the casual tourist could ever achieve.

Of course, if you're in a hurry, there is a way to shortcut this process ...

10.
Shoulder to shoulder with locals

There was blood on the floor of the Greyhound bus. We'd been delayed due to a snowstorm, and I'd passed a restless night on an iron bench beneath humming lights at the bus station in Indianapolis. My goal was Birmingham, Alabama. I was travelling alone and knew no one there. But crumpled up in my coat pocket was a ticket to a concert by the Americana musician Jason Isbell, now something of a mainstream star, but much less known in 2013.

Queueing outside the small venue, I felt how a sci fi fan must feel at a comic con. Suddenly, the names of musicians and songs that had seemed to exist only to me were now heard spoken everywhere. I stood alone, bathing in the warm twang of the Alabama accent, and easily fell into conversation with the people standing around me, some grandparents, some closer to my age. 'Noo Zeeland? You came here from Noo *Zeeland*!'

The concert was everything I had dreamed of and afterwards it seemed only natural I would be invited by a group of college students I had been chatting with to party at one of their flats. Flash forward to a hair-raising ride across the dark streets of Birmingham in one of their cars, stopping at a petrol station to load up on beer, and then arriving at a loud, buzzing student party. I could almost have been at home in New Zealand, but for the large Republican elephant on the wall.

Having arrived from liberal New York, I was suddenly surrounded by educated young people who genuinely believed Obama was the worst thing that could possibly have happened to American democracy. One earnest young man in thin glasses spoke about having just applied for a job with the US army. He still crops up on my Facebook feed, over half a decade later, standing proudly in his uniform. I think he's a sergeant now. These were not people I would naturally hang out with and yet I was comfortable there, part of the group, because we had met at an event where we had something in common.

Attending local-flavoured events is perhaps the easiest way to go from lonely traveller to being accepted and welcomed by locals. Going to events, concerts, festivals, special interest groups or niche meetups immediately puts you in an environment where you are surrounded by people who may be completely foreign to you, but with whom you share an interest, a passion, a cause or a set of values. This context immediately bridges the chasm between the local and the foreigner, and can be an immensely valuable way to connect on a deeper level with the places you visit and the people who live there.

Far from home, with your kind of people

Standing alone at that concert in Birmingham, singing my heart out to songs few friends from back home had ever heard, I was suddenly surrounded by people with whom I shared something significant. Music can represent a shared sensibility, a shared set of values. Some of the people around me had radically different politics to me. They spoke with different accents. And yet at that moment, we were united by what we had in common. We were no longer strangers.

When we travel, we often notice the differences between where we are and what we are used to. This is, after all, part of the fun! The culture shock, the strangeness, the delightful quirks, curiosities and stories. In the previous chapter, 'Blending in', we looked at how moving past this culture shock and developing a sense of comfort due to prolonged exposure to a previously strange culture will help you to better appreciate and tune in to the more subtle differences that surround you; this enables you to discover nuances and make connections that the casual tourist would never accomplish. Attending events is a shortcut to this process.

When I quit my job in Romania and started to work as a freelance writer, I realized how important it was to travel at your own pace. My first long-term trip was to Morocco, and having time to explore the country at my own pace gave me the chance to discover off-the-beaten-track places in the Sahara desert, stay with locals in Marrakech, Casablanca and Rabat, and wander the streets apparently with no purpose (like a local).

When you travel at your own pace (no matter how that compares to the pace of other travellers), you will feel comfortable to see more of the local life, get in touch with the local people, and engage in conversations and meaningful activities they may invite you to join. If you are always on the rush because of the next plane, bus, train to catch, you won't have the time and flexibility to talk and spend more time with them, and you will leave with regret. So travel at your own pace and you will have no regrets.

Luliana M, Romania

You may have only just arrived somewhere, as I had in Alabama. Your head may be spinning with culture shock, the strangeness of a place. And yet immediately you have short-circuited your 'touristness' by going directly to a group of people, in a specific context, where you fit right in. You will continue to notice the differences, of course, but suddenly you have a window into something much deeper. Locals will trust you because of what you have in common, and *you will also trust them*. You will be treated less like an outsider, and you will also *feel* less like an outsider and observe the world around you as such. You will feel a sense of comfort. It really is one of the greatest and most meaningful thrills travel can give, to be simultaneously buzzing from culture shock and also warmly welcomed as a member of the tribe.

So how do you deliberately make this happen?

Choosing the right kind of event to attend

Travellers in countries like India and Iran are frequently invited to attend weddings, eat meals and even stay over with strangers, with hospitality and the genuine interest and benevolent intention to the guest being the only goal. Being open to this possibility and saying yes (when your intuition deems it to be safe — as we discussed in Chapter 6) is really the only thing you can do to aim for this outcome, aside of course from travelling alone or, at most, in a group of two. If you haven't yet travelled in countries like this, you might think the idea of such a spontaneous invitation is unlikely. It isn't. In fact, if you travel in the right places with the right attitude, it is almost inevitable. But if you are travelling in cultures that are a little more reserved, or if you are unwilling to gamble on spontaneity and want to rig the game in your favour — or if you just want to guarantee a different kind of experience than your average tourist will have — you will need to do a little bit of planning.

The most delicate part of engineering a bit of local spontaneity is choosing the right kind of event to attend when travelling. We'll explore some ideas in a moment, but before then, let's quickly look at the kind of events to avoid. Remember, the goal here is to pursue a genuine passion of yours in order to meet locals with whom you have something in common. Look to events that are attended *by locals*. Events marketed to tourists, such as major international music festivals, self-help style retreats or tours of local sights, are unlikely to be useful for this purpose. That's not to say there's anything wrong with this — go to all the festivals and attend all the yoga retreats you like. But have a realistic expectation that if an event is marketed to tourists, then tourists are going to show up.

With that caveat in mind, let's look at the kind of events that will actually be attended by locals in a context where you as a traveller have a realistic chance to connect and become part of a 'scene'. There are probably more relevant choices open to you than you may at first realize …

Concerts

If musical gatherings of any kind are permitted where you travel, they can be a great opportunity to get to know locals on a deeper level. Of course, in addition to possible restrictions on gatherings, it's also a matter of luck if a certain artist with the right level of popularity (not too much!) will be playing in the right place at the right time. With my Jason Isbell concert, I deliberately structured my trip around travelling all those hours on the Greyhound bus to attend a concert of his in his home state, and was lucky — I'm still smug about this — that I caught him just at the start of his rise to real stardom. It also helped in this situation that Isbell's Alabama-roots style of music had extremely strong local ties, which compounded the significance of the experience. You might not be into roots music. But think about your top five favourite artists and where they are likely to have the most passionate fans. And if these places are off the beaten track and vaguely compatible with your travel plans, then consider attending!

Sports

Anthony Bourdain famously practised Brazilian jiujitsu for two hours pretty much everywhere he travelled to. He didn't write about this in detail from a travel perspective (as far as I can tell) but you can imagine that the experience would have been a fantastic way to escape the bubble and connect with locals on common ground.

Martial arts often have a certain kind of cult appeal and people who practise similar styles immediately have a lot to talk about. With team sports, the vibe is a bit different, but the after-match drinks are a fantastic chance to bond with your erstwhile teammates. Jumping on Google and seeing if there is an open football, rugby, tennis, volleyball or whatever meetup on in a city that you are passing through could afford you a fantastic opportunity to fit in with local life.

Hobbies

No matter how niche your hobby is, chances are there are other people around the world who share this passion. Do you collect stamps, play Dungeons & Dragons, enjoy improv or debate the relative merits of medieval furniture? Toastmasters events are often held in English-speaking, even in non-English speaking, countries.

Jump on Google and see if you can find a group around this hobby in one of the places you are passing through. Then *show up*. Even if it means skipping a couple of tourist highlights, you're bound to have a whole other kind of travel experience, and you might just end up making friends for life.

Bonus: networking and meetups

Travel meetups are now ubiquitous in bars across the world. While it may seem contradictory on the face of it, such meetups can often be a fantastic place to meet locals, as they often attract people who want to practise their English. I've attended dozens across Europe and Asia, and they are very easily found by searching on places like Meetup.com and Facebook events. Because they don't quite have the same 'you're one of us' magic as shared interest events like sports, hobbies and concerts, I include them here as a 'bonus' — but they should not be overlooked. One hack here is to search not just for foreigners' meetups but for English language exchange groups, where the local-to-tourist ratio will often be more promising.

***

These are just some ideas to jog your creativity. There are others that it would be odd to put on the list because you can't really plan them — being invited to a wedding, for example, is not really something you can accomplish through a Google search. All you can do is be open to these events when they do happen (and they will). In the memoirs of the former New Zealand prime minister David Lange, he describes the moving experience of attending an Alcoholics Anonymous meeting while travelling in India! There is very little limit to how much you can push this — provided you are sincerely and legitimately part of the niche group you are a part of.

No matter which option you choose, it is essential that you show up with a sincere attitude and genuinely want to immerse yourself in the experience. Don't constantly look around hoping to have a magic, spontaneous local experience. If you do this, you will most likely ruin your own enjoyment of the experience. Also, don't go with a large group. If you do so, you will inevitably seek refuge in each other's company and insulate yourselves from the experience around you. It's also much harder to invite ten people back to a house party than one or two! Take a deep breath, show up and *get into it*. Lose yourself in the music, the sport, the activity. Allow yourself to get swept away in the shared experience of bonding with total strangers from the other side of the world. And do not be surprised if this casual experience leads to invitations, lifelong friendships and opportunities to greatly expand your understanding of the world.

TRAVEL PRACTICE
The events you attend

With a notepad (digital or pen and paper) take stock of the events, gatherings and meetup groups that you have attended over the last few months. What concerts did you go to? What sport or fitness events did you participate in? What meetup groups did you attend? Go through this list and take note of three to five options you are passionate about that you can explore when you're next travelling. Then choose a destination you are planning to visit and Google these events. See what concerts are playing that fit your quirkiest, most niche tastes. See what open sports matches are on that invite participation, or drop an email to the organizer letting them know you're coming to town, and ask if they could use an extra player. Search Meetup.com, Facebook events and CouchSurfing.com, as well as good old-fashioned Google. Create a few options and then make the effort to *actually attend* when you are there!

By following this strategy, you'll give yourself the chance to stop being a tourist, and live as if you were a local, even if only for a day.

11.
Processing your travel experience

In Madrid, I lived in a garage. It had been converted, pretty convincingly, into a tiny apartment, but the metal doors that let in the sun, the harsh concrete walls, and not least of all the size left no doubt about the original purpose of this tiny rectangle. There was no desk, just a bed and a tiny beanbag on the floor. My girlfriend would leave for her job in the mornings. Soon I would get a copywriting client in town and gain the use of their office, but for some weeks I was left completely alone during the day. And so I'd wander the streets, laptop in my backpack, and work from various cafés. The time passed quickly enough.

Wandering through the Chamartin neighborhood, an affluent and only slightly obnoxiously corporate district, I'd drop anchor for a time at roadside *cervercias*. On the pavement the trendy youth and the heavily made-up middle aged would sit on rickety, hollow metal chairs, smoking cigarettes and gossiping. Inside behind the counter, the large and typically bald owner would laboriously carve slices off a leg of ham. Expensive cars rushed by outside on the clean streets, while I set up my laptop on a rickety table and powered through emails. After some hours, I'd holster my laptop and wander in search of another 'office' for the afternoon.

The 'aimless wander' is, to me, the ultimate travel experience. And at first, every street is interesting, each shop a chance to discover a new world. The students who hung out in front of the design college: dyed hair, androgynous, pierced, smoking furiously between classes. The dark suited, silent figures who flitted in and out of the Vietnamese embassy, vanishing inside behind the blacked-out windows. The tables of old men, swollen from beer, sitting on chairs groaning under their weight, somehow managing to emphatically gesture while keeping their arms firmly folded across voluptuous chests. Each scene is recorded. It becomes part of the narrative of a place. Takes your full attention. Is stored up for future reminiscence.

But then, after some days or some weeks, the level of novelty diminishes. This has its benefits. Layers of subtlety reveal themselves to you, as a result of your knowledge and ability to compare, that create a greater depth of experience. But this also has the result of nudging you, at least for some stretches, out of the physical world and into the recesses of your own mind. Wandering becomes an exercise not in observation, but in meditation. You find yourself walking through the maze of your inner world, the city you're in merely the background music to your thoughts.

It was during one such reverie that I suddenly stopped. Something that had previously been distant, liquid, evasive, had crystallized inside my mind. I rushed home. Sat on my beanbag on the floor of the converted garage. Took out my laptop and immediately began to write. The writing trance lasted a mere 30 minutes. When I came up for air, I had almost 1000 words on the page. What had been written wasn't, to my surprise, any elegant observations on Madrid or Spain, nor about my previous home, Poland. It was a very specific story covering an eccentric expat I had known over two years before in Shanghai, and his adventures in trying to start up a bar out of a fast-food restaurant in a dodgy part of town. The resulting piece ended up being published by *Roads & Kingdoms*, the travel publication started with help from Anthony Bourdain, and remains one of the pieces of travel writing that I'm most proud of. And it came out of nowhere.

Immediate clarity and insight into an experience in Shanghai had come to me, accidentally, only two years later and halfway around the world. But it was worth the wait. The culture of 'travel blogging' encourages us to immediately process our travel experience. To keep the friends, family and fans up to date *now*. But sometimes, it's more powerful to let an experience simmer, and allow it to come back to the surface only when the experience has been fully absorbed, processed and understood.

The meaning of your travel experience is not always obvious ... right away

Even if you, dear reader, do not have literary pretensions or writerly goals, there lies inside the heart of every traveller something of a storyteller. Look inside and chances are you will admit that, during both the best and worst moments of your travels so far, there is within you a stenographer furiously scribbling notes, already anticipating the moment when this experience will be processed into a story and retold to a suitably impressed audience.

This process is natural. Humans are born storytellers because stories are the engine of culture, the first draft of history and of course the origin of all religions. It also has the dissociative benefit of allowing you to step out of an immediate experience and see it as a story that you will later tell; this is a massive benefit during encounters with the delays, detentions, illnesses and inconvenience all travellers eventually encounter.

The downside of what I'll call the narrative instinct is that it presupposes that once complete, a travel experience is over. It is dead and lives on only in stories that you may, with time and the fading of memory, embellish, but will never truly stop and reinterpret. The experience is retold, but it is not relived. And after a number of retellings, you are not really remembering your travel experience at all — you're remembering your last retelling of it, putting a shadow on the stage.

The most important travel experiences we have tend

to be the most complex. That time your flight was delayed for five hours on the tarmac at Hong Kong airport is an easy one to retell. But the story of a friend you made in Havana — who interrupted the spontaneous street-side rum party you'd somehow been invited to, to take you around the corner to his house and introduce you to his mother before taking you back to the party — could take longer to process. Especially if you saw the same man a while later cycling along a busy Havana street and, in response to your question 'where are you going?' he replied, 'I'm off to send an email to my sister.'

Both of those examples are real. The first is simple: a travel nuisance story. Every traveller has a swag full of them. The second is more complex. It conveys profound differences of social attitude, wealth and lifestyle: I would never drag a new drinking buddy home at 3 a.m. to wake up my parents. But his mother seemed delighted. And if I want to email my sister, I use my smartphone. I don't have to cycle halfway across the city to a government-monitored internet café. This experience happened to me and a friend of mine back in 2013, when travelling in Havana, and I am still thinking about it, unpacking it.

The fact that it stays interesting and mysterious and challenging to you even years after it happened, is the sign of a meaningful travel experience. It's also the difference between a song by Bob Dylan and a song by Britney Spears. The Bob Dylan song may not sound so great on a first listen, but every time you play it more layers of meaning are unlocked. The Britney song sounds great and catchy, but after a couple of plays you've pretty much got it. (Not dissing Britney, I need to clarify. *Circus* is on my playlist and I'm proud. Just making a point, okay?) It's also why, despite having tried many times, I have in the end decided not to keep a travel journal.

> *I learned so much living in Paris — moving there without speaking the language was a steep learning curve in itself! But it was when I simply wandered my local neighbourhood and got lost down its tiny streets that I learned the most about how the locals lived their lives, and that was invaluable.*
>
> Karen G, Australia

The journal writer's dilemma

Journalling is in my blood. My aunt, Alyss Thomas, wrote a popular book entitled *The Journal Writer's Companion*. She and my many other well-meaning writerly elders have politely yet persistently encouraged me to journal when travelling. And I have tried. Really, I have. But for me ... it doesn't work. In fact, it is counterproductive.

There's a certain electric charge that happens when you are ready to record an experience. Sometimes this charge happens right away. Sometimes it takes two years, and hits you all of a sudden when you're thousands of miles away from the scene of the experience itself. The problem I have personally experienced with journalling is that it too quickly externalizes those experiences. This, to me, dissipates that energy and prevents this electric lightning bolt of inspiration from ever striking.

Journalling takes a thought or experience safely out of your mind and stores it on inert paper. This is why I am a rigorous list-taker in my professional life: writing down a task or idea saps the anxiety and emotional charge from it so I can get other things done (for example, sleeping) and return to it when I am ready. For this same reason I no longer keep a regular journal of my travel experiences: I want the incomplete thoughts, experiences and impressions rattling around in my head and causing as much chaos as possible. Only when they are cooked and ready do they deserve to be set to paper. I know this system has its cost — details and some experiences are doubtless lost to the ether — but it is a trade-off that I have consciously decided to make.

This is the situation for me, but having spoken to other travellers, it seems that I am an odd case. You need to experiment and discover what the situation is for you. Do notes from your travels jog your memory, or do they blunt the emotional power of an experience by making a memory seem just a little too real, as if the factual were debating with the happy narrative you had since crafted? The only way to find out is to experiment and find out what works for you. For some travellers, journalling is the ultimate key to unlocking, preserving and processing the richness of an experience. For me, that isn't the case. But you may be different.

Some use journalling to record immediate feelings about an experience. 'Dear diary, today I went to Paris, ate 96 croissants and felt happy.' Others use it to record facts. '9.37 a.m. Train number 24601 pulls into Gare du Nord. I am wearing green shoes. I take the metro to the Hotel Royal ...' A precise appreciation of detailed facts is essential if you're a journalist, a historian, a heart surgeon or a politician (just kidding with that last one). But is a rigorous commitment to recording detail really going to enhance the travel experience?

Travel is to be appreciated, not merely analyzed

It's a slightly concerning cliché that one should never 'let facts interfere with a good story'. We live, very much, in the age of bullshit where different political groups live in their own reality, each of whom assembles facts to support their preconceived notion. Hillary Clinton is a crook. Donald Trump works for the Russians. Chances are you rolled your eyes at one of those, but emphatically agreed with the other.

The fact is (ho ho) that it is, in fact, impossible to factually assemble all of the facts. In any given situation we are so overwhelmed with data that some level of filtering is essential to preserve our sanity. So even if you are a meticulous journaller, there are going to be things you choose to include and things you choose to leave out. Those things that you choose to include could form the scaffolding on which you can structure the deeper process of reflection and understanding that I rhapsodized about above. Or it can distract you from it by emphasizing analysis over feeling.

Let's take an example. Many of us get our first real experience of literature at school. I was taught Shakespeare, Steinbeck, Mansfield and various others. But I was never taught Hemingway. This, in my opinion, accounts in part for the fact that throughout my twenties Hemingway has been one of the writers I have enjoyed the most, and whose novels and short stories I keep returning to, finding in them enormous joy. Whenever I speak to a literate person and they tell me they don't like Hemingway, I always ask them if they studied Hemingway in school. The answer is always yes. Next question: which book did they make you read? Answer: *The Old Man and the Sea.*

The Old Man and the Sea is one of Hemingway's greatest works because it is the perfect distillation of his unique style. It is not my favourite book of his, by any means. Starting someone off on *The Old Man and the Sea* is like forcing someone to drink a pungent and smokey Laphroaig whisky before they have been given the chance to acclimatize their palate with a smoother, more accessible Johnnie Walker Black. What is worse, you're not even allowed to swallow and even try to enjoy it — you have to spit out the expensive whisky and analyze it with a clear head.

While schools should teach an appreciation of literature by showing people how to find joy in it, instead they teach analysis. A book is read and immediately dissected into a thousand different pieces, divided into technique and form. In the end, the student could tell you a thousand facts about every page. And yet they have learned nothing of value, because they have not been given the time and opportunity to truly feel the work — to live it. No wonder school puts people off Shakespeare!

Like a good book, a travel experience also should also be distilled emotionally. Accuracy is important, but not as important as truth. And when the earnest traveller begins deciphering their experience with the meticulous precision of a high school English student, then the experience is destroyed, not preserved. Five years later, it doesn't matter if that train was at 9.56 a.m. or 10.56 a.m. What matters is the totality of the experience. And that totality must be given time to form without excessive scrutiny or interference.

Notes can certainly be beneficial for this process. Many, if not most, of the great travel writers were assiduous notetakers. As I said, you must find the technique that works for you. But whatever approach you choose, understand that even if you are annotating every second of your day because that gives you pleasure and is a beneficial strategy for you, the goal of your notes is to make a stock-list of the raw material you have gathered, and not to create a definitive and permanent summary of your experience. Because meaningful conclusions can and must be given longer to form.

The antidote to reactionary politics is … patience

Travel taken seriously is a powerful weapon against hate. To take travel seriously means giving yourself the time and mental space to properly absorb and process your immediate experiences, allowing yourself time to form meaningful and holistic conclusions. We explored earlier how the benefits of seeing the world as a traveller can be extremely powerful when applied to your life and attitude back home. The theme of this chapter is a particularly important one to take home with you, because allowing yourself time to process experiences before drawing conclusions is one of the many things that is very much missing in today's hate-filled political landscape.

Reactionary politicians aim to oppose progress and return society to an idealized past. In the migrant debate in Europe and the Brexit debate in Britain, the reactionaries focus on foreigners: 'They are here now, and that's bad. They didn't used to be here, and that was good. Let's go back to the golden old days (but with more iPhones and less polio, please).' It's not hard to understand this perspective — it's deeply human and natural. It is also the result of a failure to take the time to properly process and understand the true implications of a situation. They see something, immediately don't like it, and call for it to be overturned. But society is going through a profound moment of change. We are in a complex situation that decries simple solutions. As has occurred before at key turning points in civilization, it is essential that people, politicians and the media be willing to make the effort to understand the historical moment that we live in. But the reactionary instinct is not interested in understanding. It does not wish to be informed. It reacts. The word 'react' implies an instant, instinctive and emotional response to an immediate situation. Once again, we see that hate is, in many ways, a failure to travel.

When Richard Nixon visited China in the 1970s and met with Mao's wiley, urbane premier Zhou Enlai, a question was asked about the impact of the French Revolution. His famous answer: 'It's too soon to tell.' This has been quoted time and time again about the wisdom of China's big picture perspective on the world, contrasted unfavourably with us hasty westerners and our tendency to leap to conclusions. Unfortunately, as reported by the *China Digital Times*, this wise remark was actually the result of a miscommunication: Nixon thought he was asking about the French Revolution of 1789.[1] Zhou, however, thought he was being asked about the student rebellion of 1968. Ah well. We have already agreed not to get too hung up on facts, right?

Apocryphal though it may be, the above story illuminates the point I am trying to put forward. The true impact of events is not always immediately obvious. Right now, the reactionaries see immigrants as a threat to their culture and safety. In England, Polish shops that crop up are seen as supposed evidence of the loss of some sort of mythical British purity. More seriously, well publicized cases of attacks by Muslim immigrants in Europe, not to mention the matter of terrorism, is more than enough evidence for some to conclude that letting these migrants and refugees into Europe is a mistake that needs to be undone. But the immediate ramifications of a situation do not necessarily equal the long-term truth of that situation. In fact, the two are usually very different.

Calm reflection shows us that migration is a fact of human history and its effects are, on the whole, positive, with migrants filling labour gaps, contributing to entrepreneurship, and creating a diverse society that is more interesting, colourful and dynamic.[2] And what of the argument that migrants are fleeing, rather than solving, problems at home? 'These young people should stay and fix things back at home, rather than coming over here' goes one popular reactionary narrative. But take a step back and talk to migrants, and you'll learn that in fact they *are* helping things at home. As explained in an op-ed during the 2015 crisis by former UN Secretary General Kofi Anan, in 2014 alone migrants sent an estimated US$436 billion in remittances back to developing countries — this is far greater than the sum of international development aid given by developed countries.[3] And if you track the root cause of the problems these migrants are fleeing — as we explored in a few examples when we examined the Axis of Evil — you'll see that, more often than not, their country did not collapse on its own. Chances are, the interference of more developed nations contributed to the crisis that sent desperate people fleeing their homes and seeking refuge in more stable lands — the very places that sowed the seeds of destruction in their own country in the first place! Talk about chickens coming home to roost.

Right now, many see the European migrant crisis as the clash of two radically different cultures: western values of liberal secularism opposed by a flood of hardline migrants intent on causing havoc and creating cultural destruction. Some fear a war between civilizations. Others see lands they perceive as Christian under attack from a 'foreign' religion (as if Christianity was born in Europe ... anyway). But what if the long-term result wasn't war? What if it was the spread of ideas between cultures? What if Europe took the opportunity to learn from visitors, as we discussed in an earlier chapter, and gained an appreciation for the human impact of their foreign policy? And what if, after an initial period of awkwardness, the migrants, persuaded by patient and rational argument, saw the merit in the hard-won values of the enlightenment, and integrated this modern and scientific world view into their own culture? This is a hopelessly naive dream. But is it really less probable than the worst-case scenario? The fact is, we don't know what the results will be. But if we are willing to be patient and learn, we may be better able to shape them for the better.

It's natural to immediately reach for emotional, instinctive conclusions. But true insight requires that things be placed in their proper context and looked at from many angles. This cannot happen instantly. When a friend of the writer Norman Mailer said, shortly after the 9/11 attacks, that he planned to write a book dealing with the event, Mailer told him to wait ten years. His words: 'it will take that long for you to make sense of it.'[4] This advice was, of course, wasted on the Bush administration, who wasted no time in entangling the American military in a series of complex foreign engagements that the United States is still trying to get out of today.

The quotidian and exceptional experiences we encounter when travelling are, hopefully, nothing on the scale of 9/11, but they too deserve their gestation period. Sometimes, it can be wise to resist the urge to immediately crystallize an experience. True insight and understanding can only come later. And by showing a willingness to wait, you protect yourself from the knee-jerk reactionary impulse that has been the cause of so much hatred and mistrust.

TRAVEL PRACTICE
Describe a half-forgotten trip
Grab your writing tools of choice, line up a motivating playlist and pour yourself a stiff drink. Then, sit down and write as directly and truthfully as possible about a significant experience you had *more than one year*

ago. It doesn't have to be a travel experience, though it can be. Ideally, make this something you haven't written about in detail before. If you have written about it, don't refer to your past writings. Don't check your journal. Don't look at your notes. Don't look at old photos to jog your memory. If you can't remember a precise detail, just guess. Allow the meaning of the experience to filter up from your subconscious during the process of the retelling. And in so doing, you might discover more insights and understanding, filtering through your current experience and matured over time, than you realized at the time or in the immediate aftermath of the experience itself.

If you aren't a writer or if this exercise gives you traumatic flashbacks to homework, then just *think it through*. Imagine yourself at a bar telling this story to close friends. How would you describe it now, and how does this compare to how you described the experience at the time? What do you choose to emphasize? Maybe at the time you would emphasize the physical — how hot it was on the bus, how hungry you were, how your friend was carsick, how the old lady with the goat on her lap dozed off on your shoulder. Maybe now you find yourself focusing more on other senses — the western pop music on the radio, the stack of hay bales outside the huts that the road passes by.

In Chapter 9, 'Blending in', we explored how staying in one place for a long time can reveal unique details that the casual visitor would not spot. Time and distance from the event itself can have the same benefit, allowing you to distill what really matters. But what if you are on the move, and determined to see beyond the surface, and return home with a unique story worth telling *now*?

12.
A story worth telling

It happened again. An old man this time. Long white hair melting into his beard. Eyes glazed over. The plump woman shouted, 'Oi! You gotta pay for that!' She ran out of the store and grabbed the man firmly by the wrist. He continued to stare blankly into the distance as she berated him. Eventually, he released the handful of goods he had pilfered, and shuffled away.

The attendant waddled back inside, her dyed black hair shining in the brightly lit petrol station shop. She looked at me and rolled her eyes.

'That's the third one tonight you know, mate. They can't stop themselves.'

'Any news about the bus?'

'Shoulda been here three hours ago.'

'I know. Did you hear anything?'

'Musta broke down or something.'

'Yes. So what happens?'

'Dunno. They'll send another. I guess.'

It was 3 a.m. in Alice Springs, in the red heart of Australia. I'd been waiting in this roadhouse for four hours, my enormous backpack propped against the chair, one leg slung through the strap, dog-eared

Lonely Planet on the plastic table. I was seventeen years old, and had no idea what I was doing.

The relentless pursuit of the unique

I guess what I was really looking for at 3 a.m. — a kid fresh outta high school now alone in the middle of nowhere Australia — was a story. I wanted to do something none of my friends ever had, to see things that others hadn't. I was impelled by curiosity, but also, I must admit, by vanity. I wanted to be *the* traveller, who had gone places and seen things that no one else ever had. This is a possible definition of travel: the relentless pursuit of the unique. And on this first trip to Australia I kept believing, time after time, that I had found it. But of course, I was never the first.

A woman walked into the store. With her battered face and yellowed eyes, it was impossible to tell her age. Intoxicated, she stumbled towards me. I tried to avoid eye contact, nervously staring at the floor.

'Hey boy. Got any bucks?'

To my relief, she soon got distracted by one of the refrigerators. She walked towards it and grabbed a cheap plastic bottle of processed orange juice. Seemingly in one smooth movement, she at once opened the bottle, took a swig, dropped onto a chair in the corner and passed out with her head on the table, spilling the orange juice, which soon started dripping noisily onto the linoleum floor.

Alone, scared, I at least comforted myself with the thought: this is an adventure. I was surely now beyond the pale of the known. A brave traveller across a new frontier. But every town I stopped in there were people who'd had similar experiences. And these people, these epic travellers, weren't grizzled yet eloquent raconteurs. In fact, most of them were pretty dull.

Dull people in interesting places

Even before I got to Alice Springs on my youthful Australian odyssey, things had started to get weird. The train from quaint and civilized Adelaide to Perth took 40-something hours and traversed the world's longest stretch of unbending railway. Out the window, you see hawks, wild horses, kangaroos, and emptiness. The richer travellers got cabins with beds. I was in a soft chair, sandwiched between two muscular female tattoo artists from Minnesota. After a few hours the conversation died. I read my book and stared out the window, hoping for the flicker or movement from a feral camel or bird of prey, but mostly just entranced by the pure nothingness of it all.

We emerged from the train with bleary eyes and shaky legs. Various touts were lined up in the train station, waving brochures for tours and hotels. Exhausted, I foolishly entrusted myself to a middle-aged Australian man called Paul, who owned a youth hostel he assured me was 'right in the centre'.

Youth hostels in Sydney or Melbourne are flash affairs replete with athletic German tourists. Outside of these cities, they get strange. In Katherine in the Northern Territory, I paid a $50 deposit for an iron key and had to chase chickens away from the dorm room. In Tennant Creek, a truly desperate town, the lone other guest was an itinerant carpenter, and the breakfast was Weet-Bix cereal without milk. In Mt Isa, a Queensland mining town, the hostel was used as an extension of the local hospital — people from the countryside travel there to await treatment in the 'city'. Various deformed figures haunted the halls at night, hooked up to beeping respirators.

But of all these places, Paul's hostel in Perth was the weirdest. Maybe because it was the start of my trip and everything was novel. Maybe it was the drowsiness from the epic train journey. Whatever it was, it just seemed that everything about Paul's hostel was wrong. The guests were Dutch tourists who lay catatonic in the garden under clouds of smoke. The vibe was less music festival, more opium den. When the weed ran out after midnight, the manager, a bald New Zealander, left the 'reception' and drunkenly shouted at the guests, demanding a joint. The water in the showers was cold. The dorm room filthy. And town centre at least a 20-minute bus ride away.

In hostels such as these across Australia, I would always meet people who had been where I had been, and seen the same things. Or, if not quite the same, who had at least had adventures of their own that were at least as epic. Some of these people would become 'hostel friends' — the type you get drunk with and confide in for a couple of nights, and hope you will never see again. Good or bad, this level of company seemed to me to dull the magic of travel. It was a cure for loneliness, but it also broke the illusion that I was alone on this vast continent that was waiting impatiently for me to discover it. Hearing about a remote town you have explored, a beautiful building, a crazy street, retold in the laconic drawl of a cashed-up 20-something on the phone to his mother while begging for more travel cash breaks the spell, somewhat. I knew that I wanted — needed — more from travel than those around me. But I did not yet have the tools, emotional or intellectual, to gain this from my surroundings. So, as many of us do, I resorted to pushing it further, taking longer journeys to stranger places, hoping to discover pure, unspoilt travel.

The best stories find you ... but you have to be willing to listen

In fevered pursuit of unique travel experiences, I rode Greyhound buses through every state and territory of Australia for a total of 20,000 kilometres, at least according to my ticket. In my mind, these buses were full of vagabonds, drifters, gangsters. Except for the tame east coast routes, tourists don't ride them. There were tattooed miners. Indigenous families. The occasional drugged-out itinerant worker from Europe or New Zealand. On a particularly long stretch somewhere far west, an elderly Aboriginal gentleman sat down next to me. He wore an Akubra hat and a smart business shirt. We exchanged pleasantries, before the rhythm of the journey took over.

The pattern is familiar. The roar of the huge bus racing at 130 kilometres per hour, punctuated by the occasional crash as yet another unfortunate kangaroo bounces off the stainless steel bull bars. We'd stop for refreshment at a lonely roadhouse, often the only manmade structure visible in the red soil. The owners would be a married couple, who were typically appalled at the idea that anyone would patronize their restaurant. It was a $3 can of Coke, and then you were on your way.

After pulling out of one such dismal establishment, the Aboriginal gentleman in the neighbouring seat tapped me sharply on the knee. 'I'll tell you a yarn mate,' he said.

'I had a mate who wanted to drive his clapped-out car right through the desert,' he said, gesturing out the window away from the coast.

'He come up to a roadhouse and he see on the blackboard outside "fish & chips", but he nowhere near the sea! So he orders it. Out come a packet of crisps, and a tin of sardines!'

We spoke for a while and the man told me something of his life. He was an engineer. He had three kids. He was travelling home after a job. His story — the fish and chips joke — wasn't great on its own merits, but in context it was, somehow, everything I had been searching for. A real connection formed, accidentally, between people from different worlds, and the exchange of a story. It wasn't an epic tale of danger or survival. It was a silly joke about fish and chips. A reminder, and a lesson to me to take myself less seriously. And also that while there may not be that many unique experiences out there, there are unique moments, unique conversations, unique connections. It's just that most of us are too caught up in ourselves to notice them.

At 3 a.m. the bus dropped me off in Exmouth, a small seaside town. 'You got a sleeping bag?' my new friend asked.

'Yep.'

'You'll be alright, mate.'

A decade later, I still vividly remember this stranger on the bus. His beard. His Akubra hat. His dark skin, crinkled eyes and constant smile. These meetings, these conversations, are often the only thing about your travel experiences that will be truly and profoundly unique to you. Anyone can hop a bus, but only you can listen, ask the right questions, and allow the magic of a conversation to take shape between people from different worlds.

The art of the unique conversation

It is the conversations you will have that will define your unique experiences. These windows into people's minds can be the result of a 'deep and meaningful' discussion, or even the product of a simple joke. But they will be your best way of gaining unique insight into a place and capturing a moment that will be treasured. There is a certain art to this, which I have gradually uncovered over many years of awkward conversations, often with people for whom English is a second language.

This relates to what we discussed in Chapter 1 on putting a place into context. If you know a little bit about where you are, this will allow you to guide your questions and also give you some insight into what may or may not be appropriate — perhaps asking about the government is a bit risky during a first interaction! Your prior knowledge can also give you clues about aspects of a person's world view and experience. But the point of doing some advance research is not to show off. While locals will often be flattered to know that you have taken the time to learn something of their country, the point is for you to listen, and if you are repeatedly bringing the conversation back to what is familiar and clear for you, you risk shutting down the interaction.

Instead of showing off, it can often be surprisingly helpful to flaunt or even exaggerate your ignorance. When talking to a stranger on a bus or in a pub, a joking admission of your own cluelessness can absolve you from any chance of giving offence (after all, how could you know better?) and allow your interlocutor the coveted opportunity to show off their knowledge and act as your guide and lecturer. This is an opportunity that I have found people young and old are more than willing to embrace under the right circumstances.

> *I took a copy of* Anna Karenina *on the Trans-Siberian Express, thinking it would be an ideal time to read this classic. The book stayed shut for the entire journey. My 'reading time' was spent either engaging with local Russians curious about why I wanted to go to Siberia or gazing out at the intriguing and captivating scenery. I learned to appreciate capturing a sense of place and that small encounters make a difference to the essence of travel writing.*
>
> Rachael R, United Kingdom

Sometimes, these questions get a bit uncomfortable. I have spoken with people from the Baltics or Russia who have passionately ranted about how homosexuality is a poison that is the result of American propaganda. I find this point of view distasteful, and of course it is nonsensical. However, it is unfortunately the way many people out there see the world. Taking offence, ending the conversation or resorting to argument is unlikely to change views. Instead, view this conversation for the unique and interesting travel experience that it is, and take it as an opportunity to learn about how others see the world. Genuinely enquire, as if from a position of benign, ignorant neutrality, as to how this person arrived at this point of view. 'How do you … ' sounds combative. Instead, you could ask, 'What shows you that this is true?' or, if in doubt, simply, 'Can you help me understand more about this?'

'But wait a minute, Nathan,' I hear you cry. 'You're saying it's my perspective that makes my travel experience unique. But you're asking me not to have a perspective when talking to people?' Well, yes and no. Of course you will have a perspective on issues, events and ideas — it's your right as a human to do so. But think about the previous chapter on processing travel experiences. If, for a beautiful moment, you can put your own world view on the back seat and act as if you fully absorb the perspective of someone else, you will expose yourself to much more data, much more material, than you otherwise would. You can then reflect and process this through the lens of your own perspective, resulting in a glorious cocktail of understanding and insight that will make you the envy of any youth hostel prophet from Sydney to Seattle.

If you're too focused on having unique travel experiences, chances are you won't have any. You'll be too wrapped up in comparing yourself to others, too eager to instantly process your experience into a retellable story, rather than allowing it to properly form in the back of your mind. Instead of holding out for unique experiences, start creating them, not through your actions but through your perspective. This is, in part, what this book has been arming you to do. Anyone can sit for twelve uncomfortable hours on a bus and wander about in a strange town. Going there alone is nothing special. It's *how* you see it that distinguishes you as a traveller. Most importantly, your perspective can guide your curiosity. And at its best, this can cause you to *listen* — to be open to the chance meeting, the surprise conversation, that you will end up treasuring for life. This way, when you do finally get that moment of magic on a train in the middle of Siberia, on a boat in the Amazon, or even in a quiet corner of your home town, you will be equipped to do justice to the experience and immerse yourself completely into it, seeing it for what it is, and finding the core, true uniqueness that made itself visible only to you.

TRAVEL PRACTICE
Talk to someone you strongly disagree with
Much of the advice in this book centres around the core idea of applying the openness and curiosity we grant to strangers when we travel to those we encounter every day back at home. Let's take this idea to its logical extreme. Imagine you are David Attenborough observing a crocodile devouring a poor lamb, but you have resolved not to intervene. Imagine you are an intrepid journalist who has been let into the opulent office of a murderous foreign dictator, and you have to listen with patience and courtesy to his inane ramblings lest you leave with fewer limbs than you brought with you. Get yourself into this state of non-judgmental receptivity and call up your racist

uncle. Or that Trump voter you know. Your grandma who voted for Brexit. Your cousin who is opposed to gay marriage.

When you find someone with whom you strongly disagree, and speak to them about that very issue that divides you, do not attempt to persuade them. Suppress your revulsion. Simply *listen*. No, you are almost definitely not going to change your mind as a result of this conversation, and they aren't going to change theirs, but there may be *accidental* results that are equally positive. You might hear a story you've never heard before. You may discover something, some tiny key to this alien perspective, that helps to inform and add texture to your own world view. Or you may not. The story isn't always there. But you won't find it if you keep getting in your own way. This is an exercise in extreme receptivity; with practice it will serve you well both at home and abroad.

13.
Finding your magic

It took Jack Kerouac just three weeks to write what is possibly my favourite book of all time: *On the Road*. Fuelled by alcohol and Benzedrine, Kerouac wrote the entire novel on one huge scroll of paper that, when unrolled, measured 9 metres (30 ft) in length. The novel was based on the author's travels across the United States and (briefly) Mexico, in the company of various eccentrics including Allen Ginsberg, Lawrence Ferlinghetti and Kerouac's muse Neil Cassidy, referred to in the book as Dean Moriarty.

Kerouac's spontaneous prose feels like jazz. When reading it you find your head bobbing along as if to music. And then you stop seeing the words, and you find yourself in a battered car, roaring across America. You hitchhike across the train tracks and lose yourself in the rhythm of the band at a black bar in New Orleans. You anxiously arrive in Denver, looking for a long-lost girlfriend, scrounging for a dollar to put some food in your belly. You get harassed by cops who think your car is stolen, wreck a borrowed Cadillac, and watch a young prostitute weep into her beer at a brothel in Mexico.

Many in the literary establishment dismissed

Kerouac. Many more would certainly argue that his novels do not constitute 'travel writing'. But nowhere else will you find such a strong, clear sense of the power of the road. It is a book about travelling as a quasi-religious, even spiritual experience. Everything you could ever want is out there. You travel to escape yourself and the drudgery of your normal life and to find something real, to meet mad people 'desirous of everything at the same time, the ones who never yawn or say a commonplace thing, but burn, burn, burn like fabulous yellow roman candles exploding like spiders across the stars'.

To me, the book is like a drug. Stronger than whisky. You can analyze the romanticism, break it down, dissect, place it 'in context' in the same manner we explored with countries in this book. But that is not the point. It is not a book to be thought about. It is a book to be felt. It taps into that vein of wanderlust that exists inside the soul of every traveller but sometimes falls asleep, drowned out by the anxieties and drudgeries of real life. When I reread Kerouac or listen yet again to the audiobook of *On the Road* while waiting alone at some strange train station or leaky bus stop, suddenly all the delayed flights, grumpy border guards, credit card bills, jetlag, hangovers, frustrations and agonies of the road become worth it. The magic takes over and sweeps me away because, thanks to *On the Road*, I have remembered what travel is.

The American travel writer Paul Theroux wrote, 'I

have seldom heard a train go by and not wished I was on it'. For him, it was the promise of a train. For me, it's reading Kerouac, looking at the names of strange cities on maps, listening to songs by folks like Steve Earle and Townes Van Zandt about wanderers and vagabonds. What is it for you? What gets your heart racing and causes you to be overtaken by dreams of travel? It's important to remember this and tap into it, because the world will constantly try to take it from you, to stamp out the flame and make you as normal and dull and sedentary and content as everybody else.

There are those moments that happen to all of us when we forget why we are doing it. We worry about getting to the airport on time. About paying for the tickets. About our visa status. And the Airbnb booking — did we check the reviews, and what day is check-out? We arrive somewhere tired and grumpy, and instead of being lost in the thrill and excitement of the place, we start jonesing for a wi-fi connection. Thoughts of the world back at home, of obligations and anxieties, crowd our head. We engage with what E.M. Forster called the 'world of telegrams and anger' and let the place where we actually are fade into the background. Practicalities and pains force everything else to the background. Deflated, exhausted, you ask yourself, 'What the hell am I doing here?'

We all have these moments. The world is designed to keep us connected. Our brains are wired to search for threats, the 'Did I leave the oven on?' instinct. A little bit of this kind of thinking is essential — you do, after all, need to engage in some element of planning to actually function as a human being, get off your damn sofa and get anywhere. But too much is destructive. While alcohol, mindfulness and exercise are all useful tools to keep this manner of thinking under control, recapturing the magic of travel is the purest and most real sense of motivation. You will get those tickets booked, fight your way to the front of any embassy queue and get those immunization jabs with a smile if you know what it is in the name of. But when you forget, you need a way of bringing it back to the surface.

The commodification of travel as a break from our normal lives has meant that many of us have basically forgotten how to do it. We are so used to functioning in 'normal life' without taking an excessive interest in the world around us except as it relates to our own interests and ambitions, that the curiosity switch is rusty. We are so used to seeing the working day as a series of duties to be efficiently dispatched that we have forgotten how to indulge in the promise of the moment. We are constantly exhorted to make plans and set goals that we have lost the ability to be spontaneous, say yes to a random opportunity, and gleefully end up somewhere we never expected to be. Tour companies have attempted to artificially package 'travel', to bottle the magic. But real travel cannot be planned. It cannot be scripted. It cannot be 'designed'. It happens when you allow it to. And in order to allow it to, you need to get *high*, high on the promise of the world and the lust for adventure and discovery.

Once you have tapped into this magic, there's little you can do. If you aren't travelling it will eat you up inside. So you have to go and *find travel* wherever you are. We have talked often in this book about how the principles and attitudes of travel can still endure when at home. So too can this magic. While at university in the isolated South Island town of Dunedin at the bottom of New Zealand, miles from anywhere, I channelled this need for travel into long night-time drives in my 1989 Honda Concerto. Often in the company of a similarly insomniac friend, we would show up at 3 a.m. in small seaside towns, sharing the road with trucks, sheep and occasionally penguins, feeling as if the whole world were ours.

Travel exists anywhere and everywhere. You cannot pretend that you don't hear the call.

When you hear it, you can feel it everywhere. You can get the thrill of adventure when lining up in the company of exotic looking locals at a supermarket in the residential part of a strange town, or from riding a broken-down bus through the mean parts of a big city. You can feel it alone, or in company, at tourist traps and on the edge of undiscovered countries. But when you lose it, you might need to go searching further to find what really excites you. If jostling with tours and tourists at the Eiffel Tower is no longer giving you a rush, then make it your mission to discover your magic. You could push it further; go to Moldova or Kyrgyzstan instead of Italy or Turkey, and feel, as I did many years ago in Ukraine, that you're a pioneer discovering a new land. You could decide to make the next trip alone, so you can more sharply feel the travel rush. You could take the overnight train, instead of flying. You could decide to do something *bonkers*.

My phone died during a road trip through the Pacific Northwest. My partner and I stopped in the small town of Aberdeen, famous for being the home town of Kurt Cobain. Normally I would have exhaustively researched a place to have lunch, relying on ratings. Without a working phone, we were clueless.

Then I spotted a little bookstore across the street. It had funky wallpaper and a plastic unicorn head in the window. 'Let's ask them!' I said. It was owned by an enthusiastic psychedelic couple who happened to be huge foodies. We were directed to an incredible Salvadorian restaurant where I tried my first papusa, and their coffee recommendation led us to a chance encounter with a local who was planning a backpacking trip to Ireland, where we live.

So my new method of 'food finding' when I travel is to find a cool local shop and ask people where they like to eat.

Erin S, United States

In 1933, Patrick Leigh Fermor was expelled from school after he was caught holding hands with a local shop-keeper's daughter. At a loss for what to do next, the young man hit upon a bold idea. He caught the boat from England to the Netherlands and set out on a walk. He walked from the Hook of Holland to Istanbul, a city he'd always romantically thought of as Constantinople. The start of the journey was rough. He slept by the side of the road and alongside livestock in barns. But then, by the grace of a few chance encounters, he found himself the guest of a series of aristocrats who shipped him from castle to castle on a tour of the last generation of real aristocracy in Europe. Many decades later, he wrote about this in a series of books that began with *A Time of Gifts*.

Having fought in World War II and lived a rich life in the many years since his youthful adventures, Fermor's writing is heavy with nostalgia for a lost world, but also light with the joy of discovery. He had that same forceful, irresistible energy of wanderlust that so excited me about *On the Road*, a passion that shines through in Fermor's writing and that he never lost, even deep into his old age. 'I lay in one of those protracted moments of rapture which scatter this journey like asterisks. A little more, I felt, and I would have gone up like a rocket.' Kerouac and Fermor are seldom mentioned in the same sentence — the brash American beat and the polyglot British scholar and playwright — and yet the work of both positively rings with this same sense of *travel*, of the call of the road.

Unlike Fermor, Kerouac's life was a sad one. Literary success came to him too late and in too strong a dose. Unable to cope with fame, stung by what he saw as misinterpretation of his work, and hounded by hippies far younger than him who idolized the fictionalized version of himself he portrayed in *On the Road*, he drank himself to death at the age of 47. His hero, muse and greatest friend, Neil Cassidy, died the year before at 41. Their fevered dreams of travel may have been too much. Too drunk on freedom. Too naive to survive.

Fermor, an equally romantic soul, smoked up to 100 cigarettes a day and yet made it to the age of 96. He was writing until his last days. While Kerouac was fuelled by alternating euphoria and bitter depression, Paddy was possessed by a constant scholarly enthusiasm. He mastered not only several European languages but also many obscure regional dialects, tracking down historical connections and mysterious literary allusions and staying late into the night to discuss his findings with the finest minds in Europe.

To keep the *flame*, the *magic*, the excitement alive is a matter for both the heart and the mind. You can survive with Kerouac's drunk passion only if it is combined with Fermor's curiosity and interest in the outside world. Curiosity is self-refuelling. No matter whether your interest lies in literature, geography, botany or beekeeping, there are always going to be limitless horizons for discovery. In this book, we have talked extensively about how travel can be understood intellectually and how curiosity can be applied to enrich the experience and enhance the understanding of the world. When this is combined with the emotional heart of travel, the ever beating and irresistible passion, some kind of symphony is created, expanding beyond the immediate journey and radiating through every aspect of your life, bringing you closer to the world and the world closer to you. This isn't going to happen every day. But when it does, just be sure you're ready for it.

TRAVEL PRACTICE
Keep the flame of wanderlust alive

The French decadent poet Charles Baudelaire wrote: 'So as not to feel Time's horrible burden that breaks your shoulders and bows you down, you must get drunk without ceasing. But what with? With wine, with poetry, or with virtue, as you choose. But get drunk.'

What gets you drunk on the idea of travel? Is it a song, a book, an idea? Is it staring at a map and fantasizing about all of the exotically named cities? Is it looking at photos of far-flung destinations? Is it the foreign food, the music, the language, the postage stamps!? You would not have picked up this book if wanderlust — from the German *wandern*, to hike, and *lust*, desire — wasn't a part of your character. The fact that you feel the pull of the road is a gift and a privilege, but it is not something you can take for granted. So many fervent wanderers allow the flame to die, and retire to the armchair, the bottle or the TV set. If you are serious about being a *traveller* then this is a lifelong commitment, and a passion that you have to fuel, nurture and preserve.

Distil and connect with the thoughts, sounds and ideas that thrill you, that make you want to head out to the airport, look at the departure board and buy a ticket to the weirdest looking destination that you see. This wanderlust is not a cliché. It's not a Facebook meme. It's something real and it is uniquely yours. And the world will try to take it away from you. The costs and vicissitudes of travel. The discomforts of the road. The anxieties and obligations of everyday life. All of these forces and more will conspire to blunt the edge of your wanderlust. This is why you need to work to fan the flame. To indulge in the fantasy. To listen to the songs, read the books, look at the photos and think the thoughts that get that fire roaring once again. No one can keep this alive but yourself.

Conclusion:
Blurred lines

It had been more than two years since I'd set foot on home soil. Homesickness has never been much of a problem for me; a hostel room or Airbnb normally starts feeling like home as soon as the suitcase is unzipped. But two years — that was a bit too long. So being back in Auckland, New Zealand, felt good ... but also, not quite as good as I expected. There was one moment, sitting in a café not far from where I grew up, when the waiter 'kindly' asked 'What part of England are you from?' Having lived for the better part of five years among mostly non-native English-speaking company, I'd neutralized my accent. This was home, but here I was being welcomed as a foreigner. And that, too, was how I saw the place. Everything was familiar ... yet also, nothing was. My instinct was to wander, discover, seek out the weird and interesting, the same way you do on the road. The feeling was unsettling. I couldn't wait to get away again. It wouldn't take long.

Interest and circumstance blew me to Budapest, where the majority of this book was written. It has been the familiar expat pattern: wrestling with the language, getting a sense of how to get from A to B, developing a circle of local and international friends, finding cheap places to eat and local places to drink. I've also been taking my own advice, and diving into the history. I'd known something of it before visiting, but reading about a place that you're in is a great way to build excitement for your new home. The story becomes real when you read about battles fought around the corner from your apartment. The scars in the wall become bullet holes. You know the names of the soldiers who may have fired them. You get a sense of walking the set of a legend, you feel the current of the story flow over you. You're part of it. It's not just academic.

Often, as any normal resident of any normal city does, I find myself visiting the local shopping mall to stock up on socks, groceries and other dull but useful things. Outside the glistening mall filled with tattooed, pierced and either impressively fit or horrifyingly rotund locals is an old yellow building; the Corvin Cinema. The paint is darkened by decades of soot and smog. Outside it are several plaques and a few heroic looking statues. I had passed by often, but seldom paused for a closer look; the plaques were in Hungarian, a basically impenetrable language, and I had come in the 'errand' frame of mind, with a dull job to do.

After the perils of finding an apartment (an

outrageously dodgy exercise in this city) were complete, I started collecting books on Hungarian history and by Hungarian authors. While reading Victor Sebestyens's fantastic book on the 1956 revolution, *Twelve Days: Revolution 56*, I made an interesting discovery — no doubt already known to all locals but not immediately obvious to tourists. It would forever change my visits to the mall, and how I thought about that faded yellow building that stands right outside it.

In *Twelve Days*, the author describes how the Hungarian rebels had created spontaneous strongholds throughout the city, building makeshift forts from which they took potshots at Russian tanks and tried to stay alive just one more day. The biggest and arguably most significant fort? Corvin Cinema. Here, children as young as fourteen had grabbed guns and, using a network of underground tunnels, caused serious damage to Russian troops, kidnapping several tanks and keeping the morale of the revolution alive. The most successful stronghold of the revolution, it kept alive the dream of freedom and led some Hungarians to believe that they may actually be able to rid themselves of their Russian oppressors.

This is the very same yellow building I pass by day after day. Suddenly, it took on meaning. I didn't just see a faded yellow shell. While walking by, I saw the rebel soldiers in the windows, imagining the tanks rattling down the road that was now packed with cars and trams. In old photos, things don't look too different from how they do today, except for the enormous modern mall, of course! Now, every trip to the mall is enriched, somewhat, by my understanding of what happened here. A dull errand is also a pilgrimage, of thoughts, a chance to put myself into the moment of history and imagine how events looked and felt when Hungary leapt to its feet and yelled for freedom. Today, the building still functions as a cinema. If you Google 'Corvin Cinema' you get a list of movie times in Hungarian. Unfortunately for the traveller, most international films are dubbed.

These days, it's hard for me to tell when I am travelling and when I am not. I feel a stranger when I return to the land of my birth, and get a kick out of daily tasks like visiting the mall in my adopted city. When we better immerse ourselves in the travel experience we can better understand and connect with different people and different places, and so become part of the resistance to nationalism, hatred and bigotry. We are better people when we travel. We ask more questions. We think more. We question our assumptions. Surely these qualities are also ones best taken back with us as souvenirs, habits to be trained, nurtured and kept alive in all places. Travel doesn't have to happen far from home. It can happen in your suburb. Travel is not something you *do*. It's a state of mind. It's an attitude of life. And it is

something that we can once again take seriously,
invest our whole hearts and minds in, and take with
us not only everywhere we go, but also everywhere
we stay.

Acknowledgments

This book arose out of conversations between Exisle's Gareth St John Thomas and Anouska Jones. If it weren't for them, this book would never have happened. Anouska kept a close eye on early versions of the manuscript as it developed, before the draft was handed over to Karen Gee, who helped hew this into shape — thank you for dealing with my scattershot grammar and for letting me know in those (many) instances where my sense of humour did not quite translate onto the page!

Of course, when it comes to the soul of this book I have to thank my mother, Gail, for allowing me at the age of seventeen to set out alone on a trip that, it turns out, still has not ended. You understood why I had to go, and have visited me in my various homes ever since. My grandfather, Anthony Levine, ignited my passion for travel writing with early gifts of the works of Paul Theroux, Patrick Leigh Fermor, and many others. My aunt, Alyss Thomas, took me seriously many years ago when I said I wanted to write, and continues to help me understand where the words really come from.

I have been privileged to travel in the company of many wonderful friends who know who they are — may the adventures ever continue. Thank you also to those writers of Intrepid Times who gave permission for your words to be quoted in this book, and to our ever-growing community of wanderers.

Thank you to Jennifer Roberts for your editor's eye and helping to keep this book relevant in the strange post-ish pandemic reality we are now living in. And of course, to Joanna for being by my side as every word was written, rewritten, and so on.

Endnotes

Chapter 1
1. https://www.jstor.org/
   stable/4330418?seq=1#page_scan_tab_contents
2. https://www.thenational.ae/opinion/trump-s-rogue-regimes-
   un-speech-recalls-bush-s-axis-of-evil-1.661816
3. https://www.britannica.com/event/Iranian-Revolution
4. http://iran.sa.utoronto.ca/coup/web_files/markcoup.html
5. http://www3.weforum.org/docs/WEF_GGGR_2017.pdf
6. https://www.ncbi.nlm.nih.gov/pmc/articles/PMC5936844/
7. https://www.unfpa.org/news/adult-education-offers-new-
   opportunities-and-options-iranian-women
8. https://www.brookings.edu/blog/
   order-from-chaos/2019/03/14/irans-economy-40-years-aft
9. https://www.bbc.co.uk/history/recent/iraq/britain_iraq_03.
   shtml
10. https://www.globalpolicy.org/iraq-conflict-the-historical-
    background-/us-and-british-support-for-huss-regime.html
11. https://www.usip.org/publications/2019/03/
    current-situation-iraq
12. https://www.pri.org/stories/2018-03-12/
    fifteen-years-after-us-entered-iraq-baghdad-breathes-new-life

Chapter 2
1. https://www.forbes.com/sites/
   laurabegleybloom/2019/06/04/youngest-woman-person-to-
   travel-every-country/#6162a3cd4aea

Chapter 3
1. https://www.historytoday.com/archive/months-past/
   treaty-westphalia
2. https://courses.lumenlearning.com/suny-hccc-worldhistory/
   chapter/introduction-to-nation-states/

3. https://www.theguardian.com/society/2018/mar/01/how-americas-identity-politics-went-from-inclusion-to-division

Chapter 5

1. https://ec.europa.eu/eurostat/statistics-explained/index.php/Asylum_statistics#Age_and_gender_of_first-time_applicants

2. https://www.google.com/url?q=https://www.nytimes.com/2015/12/20/world/europe/norway-offers-migrants-a-lesson-in-how-to-treat-women.html&sa=D&ust=15673522143
71000&usg=AFQjCNG_Ilp7hG5QZjAn-xrcKSlnGFk0Hw

3. https://www.psypost.org/2013/12/new-study-confirms-mark-twains-saying-travel-is-fatal-to-prejudice-21662

4. https://www.independent.co.uk/news/science/white-people-become-less-racist-just-by-moving-to-more-diverse-areas-study-finds-9166506.html

Chapter 8

1. https://www.nytimes.com/2019/06/03/travel/traveling-climate-change.html

2. https://www.forbes.com/sites/jamesasquith/2020/01/13/the-spread-of-flight-shame-in-europe-is-greta-thunberg-the-reason-why/#79740aca69bd

3. https://www.nap.edu/read/12794/chapter/5

4. https://www.theguardian.com/lifeandstyle/2019/jun/14/the-mindfulness-conspiracy-capitalist-spirituality

Chapter 9

1. Rosling, H, 2018, *Factfulness: Ten reasons we're wrong about the world — and why things are better than you think*, Flatiron Books, NY.

Chapter 11

1. https://chinadigitaltimes.net/2011/06/zhou-enlais-caution-lost-in-translation/

2. https://www.wsj.com/articles/SB10001424052970204466004577102120349374652

3. https://www.project-syndicate.org/commentary/human-migration-reality-by-kofi-a-annan-2015-06

4. https://www.reuters.com/article/sept11-books/rpt-writers-take-time-to-absorb-sept-11-impact-idUSN1E7851712011090

Index